Airs

Profit and Plunder Under Sail

An Illustrated History
Written and illustrated
by:

Captain Roger C Horton

Introduction

A great number of books have been written about boats and ships, most of them excellent. In fact, there are books written about single ships that are longer than this one. My purpose in writing another book on the world's vessels was to tie their evolution to history and human development. I wished to describe the part they played in war, trade, migration and of course, pure adventure. Why did certain vessels evolve; how did they influence human history and how did history influence them.

This book separates vessels through time and by their use, by construction and world region. I have tried to show what it was like for the crew aboard vessels, and what difficulties and perils they faced. This book was created to let you understand the vessels and the men who put to sea in them. To grasp the advance of the marine development of vessels driven by wind and muscle. How we got to where we are from the moment when the first of our ancestors straddled a log to cross a river.

It is my sincere hope that you will enjoy the book.

Profit and Plunder Under Sail

Other titles by the author
To Stalk the Hydra
Racing for Pride and Profit
Truth and other Precious Things
Walking to Mombasa

Hurricane Road series.
1   Hurricane Road
2   Florida Straits
3   Valuable Things
4   A Change of Times
5   Lies and to the wall
6   Implausible Deniability
7   Paths of War

Table of Contents

Preface: How we became mariners.
The Age of the Clipper Ship.
The Raiders
The Privateers
The Postal Packet Vessels
The Wreckers
The Pilot Boats
Fishing Vessels on the Grand Banks
The Smugglers
Cargo Schooners, Tramps and Traders
The Cargo Ships of the Indian Ocean
The Vessels of the Pacific Ocean
The Iron Windjammer
About the Author

Preface

Three Log Raft Canoe

How We Became Mariners

Human seafaring is rather ancient. Archaeologists are convinced that the settlement of Australia and the Americas, beginning 50,000 years ago would have required some use of crude watercraft assisting Paleolithic migrations along the coastlines and across bodies of water, even with the far lower sea-levels.

Archaeologists have studied the evidence of an early occupation of what would become the city of Ur, during the Ubaid period, 6500 to 3800 BC). Eventually they hope to find older ships than Ubaid reed ships of Mesopotamia and Arabia which at present are the oldest found. In the delta of the Tigres and Euphrates Rivers, the use of these seagoing watercraft made of bundled reeds, coated with Bitumen was widespread. Some remnants were carbon dated before 8,000 B C. and these craft were also used in coastal trade past 3,000 B C

. Small reed boats and boats based on carved models, similar in form to Bellums are still used today on the Persian Gulf. They are little planked canoe- shaped craft, with upturned decorated bows the design going back 8,500 years.

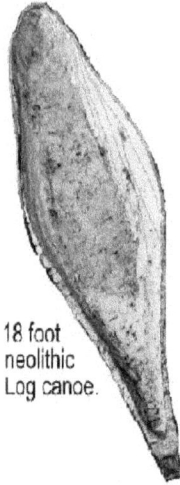

18 foot
neolithic
Log canoe.

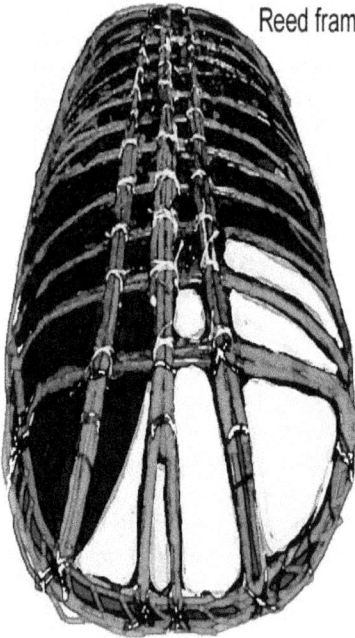

Reed frame

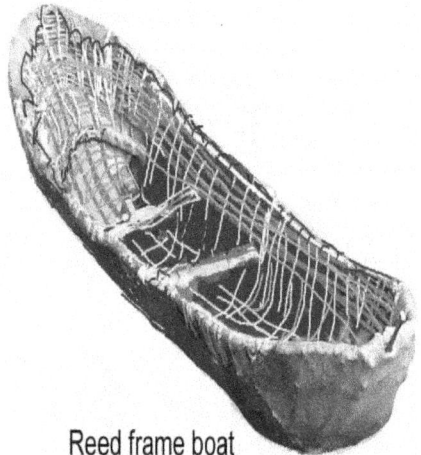

Reed frame boat
covered with hides.

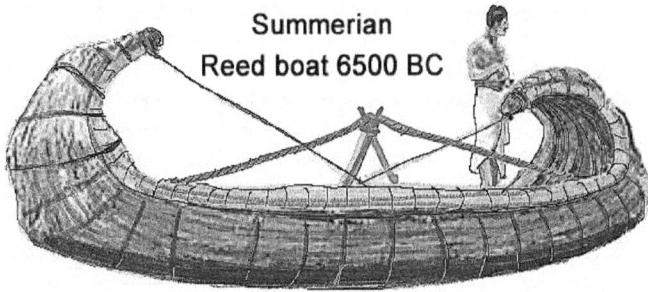

Summerian
Reed boat 6500 BC

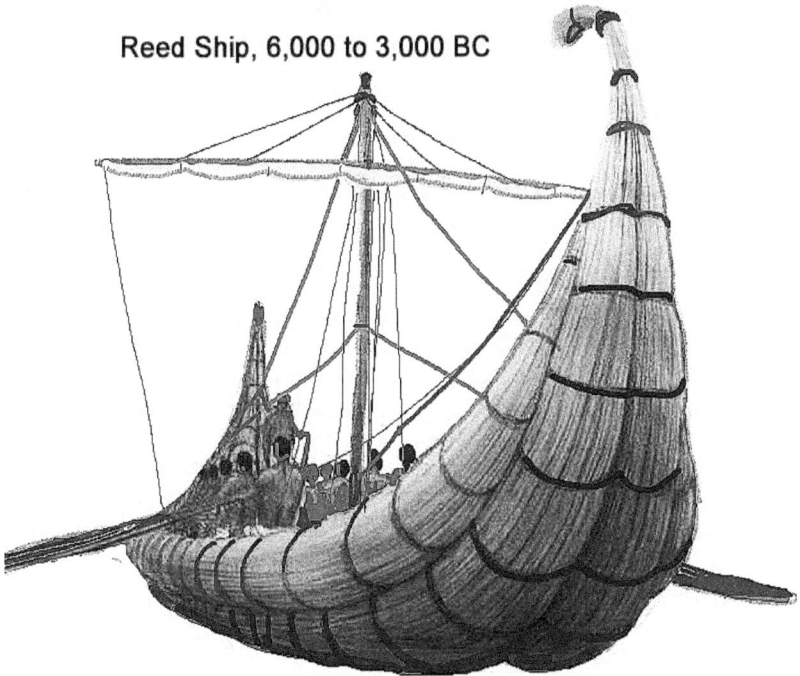

Reed Ship, 6,000 to 3,000 BC

No doubt, many millennia before any written record, even carvings or base-relief sculpture, humans needed ways to cross water. To cross rivers and lakes, or to fish from. First were logs, even rafts, hollowed log canoes, inflated animal skins, even boats with skins stretched over a wood frame. In many parts of the world, examples of small crude watercraft predate 10,000 BC. The first archaeological evidence that I'm aware of, which shows humans were crossing open sea in boats is in the late Pleistocene.

Dating before 9,500 B C, is a rock dwelling on the Island of Cyprus where tools and the cut and burnt bones of African pigmy hippopotamus and other animals were found. These animals were indigenous to the island, but humans were not. These animals have been extinct since the time of human arrival. Even with the lower sea levels of that time, Cyprus was 38 miles over open water from the coast of the mainland. It is possible that wooden boats sculpted with stone tools and fire, held together with wood pins and stitched with fibers, were navigating the eastern Mediterranean Sea, well over 10,000 years ago --- the islands of the Aegean were settled after all. The only remnants of boats found, have been simple small craft and log canoes buried in the clay and silt of ancient lakes and marshes. Vessels able to cross open oceans is an entirely different thing.

By 7,000 BC the Mother Goddess and Sacred Bull God of the Anatolian Neolithic culture spanned an area from the headwaters of the Tigres and Euripides rivers to Egypt and west toward the

Aegean Sea. This included the islands of Cyprus and Crete.

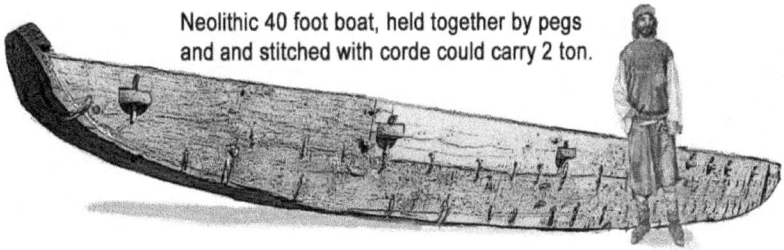

Neolithic 40 foot boat, held together by pegs and and stitched with corde could carry 2 ton.

The ancient settlement of Knossos on the north central coast of Crete is thought to be where the Minoan civilization originated. Crete, like Cyprus was uninhabited by humans before 9,500 BC. Absolutely no trace of human indigenous development exists on Crete, so, this is indisputably a case of the island being colonized by sea. By 7,000 B C they were bringing large domesticated animals, cows, sheep, goats to islands of the Aegean and to Crete. The boats they crossed the water in, would have been built of wood and sturdy. They needed to carry a lot of weight without being swamped by waves. There's no question that their domesticated animals were transported to Crete because the is no record, fossil or otherwise of the animal's existence there before the human settlement of Crete. No doubt, these were not safe ocean- going vessels but with a weather eye they made it.

The Neolithic era in Crete covers the period from 6000 B.C. to 2800 B.C. When the first men landed on the island it was densely forested by giant 140-foot high Cyprus trees, with trunks of substantial diameter. The wood is soft enough to cut easily, moderately strong and not subject to rot or insects. Best of all its very light and unless a swamped boat was heavily loaded it was not going to sink. It could be cut and worked with stone axes and other Neolithic tools. The Aegean Sea has thousands of islands and one can always see the next bit of land on the horizon but the region is cursed with spates of nasty weather. This was a far greater challenge for the ancient navigator than the modern. Given the difficulty of the marine environment there was certainly a motivation for boatbuilding innovation.

What really allowed the great advances in shipbuilding was the advances in metallurgy. The mastery of hardening copper by alloying it with other metals, particularly tin, and casting it, (the Bronze age) allowed the creation of efficient woodworking tools. The adze, drill, axe, chisel and very important, the saw. The saw allowed for the long regular planks needed to build ships. Logs could be split and shaved Scandinavian style but the saw was far faster and efficient with cedar. Now, you had efficiently built wooden ships, planks on a frame and keel.

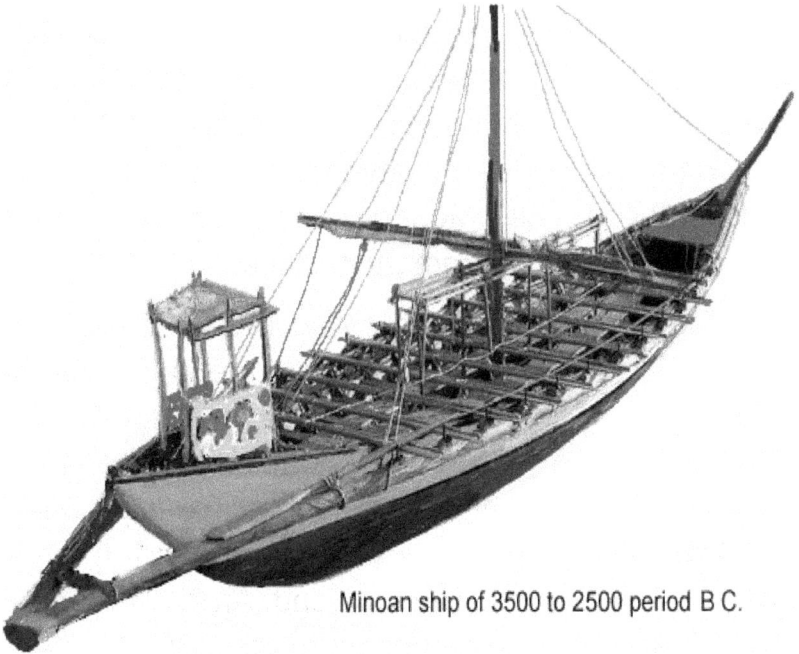

Minoan ship of 3500 to 2500 period B C.

By 3,500 B C the tentacles of exploration and trade were extending out from the Minoan kingdoms of the Aegean all the way to Iberia and beyond into the Atlantic. Shipbuilding technology spread, and advanced to Egypt, into the Red and Arabian seas, the Indian and Pacific Oceans. Trade cities and colonies were being established.

Products and ideas spread over the world's seas faster than any time in human history due to ships. Now there was the realm of the sea and the land. Not for another 7,000 years would the realm of the sky be added, and then, within 65 years, the realm of space. The Mesopotamians and Egyptians, with the galleys and barges, plied the Euphrates and Nile River; Minoans, Mycenae, Greeks, Phoenicians, Carthaginians and Romans crisscrossed the Mediterranean on the great voyages, even circummuring Africa.

Seafaring Austronesians sailed the waters of the Pacific thousands of years before the unification of China; a branch of these people known as Polynesians, explored and colonized the islands of the vast Pacific, The Ming Chinese Treasure Fleet of Admiral Zheng He surveyed and charted large parts of the world. Persian and Arab merchants traded from the African coast to the Sea of Japan, while the Celts and Nordic peoples explored the river systems of Europe and western Asia, and ventured out upon the icy edges of the North Atlantic to colonize and raid all the way to North America. Chinese, Arab, Portuguese, Dutch, English and Spanish merchant seamen, as those ancients before, sailed off the charts. The trade routes they forged were followed by conquerors and colonist as new empires were built. Some men sailed for trade, some for conquest and plunder. Most sailed for both. Such is the nature of man, past and present. I doubt it will be different when our species voyages among the stars.

Seamen through the ages, sailing craft of every description, braved the oceans elements and the unknown shores, for curiosity, profit, and glory. Each stood on the shoulders of those intrepid voyageurs who'd went before and drew charts for those who would follow. They penned in islands, reefs and passes where before had been only the words, "*Beyond here there be dragons.*"

Gradually the ocean seas of our globe were given shape by the pilots notes and the cartographer's skill. Sailing ships grew in both size and strength, speed and efficiency, reaching their pinnacle with the sleek wooden clippers and after them, the great iron windjammers whose careers stretched into the 20th century. The age of the engine, and the gigantic steel liners, freighters, tankers and warships that engines power, has driven commercial sail from the sea but what brute engine has a heart. The sailing ships and the men who crewed them, they were the valiant ones.

Sails and masts of a square rigged Barque

1. Course
2. Lower topsail
3. Upper topsail
4. Lower topgallant
5. Upper topgallant
6. Royal
7. Jigger topsail
8. Spanker
S. Stay sail
J. Jib
11. Crojack

The names of square sails remain the same as they go up but also take on the name of a specific mast. For instance, Lower Main topsail, lower mizzen topsail, or lower fore topsail. The Crojack, # 11 is the exception here.

The Age of the Clipper Ship

The first of the ships that came to be called clippers were built on the East coast of the United States. Some say that the first was John Griffith's 154-foot *Rainbow*, launched in New York in 1845 but in truth it was the 192-foot *Sea Witch* that he saw built one year later. Planned for the China trade, her rigging was extremely strong. Heavily sparred and built with especially tall masts for a vessel of her size, she had at the time of her launching the tallest masts afloat and set the record, Hong Kong to New York in 77 days.

It remained unmatched for a hundred and fifty-one years. Four years later Donald Mackay's 210-foot *Flying Cloud* was launched and proved to be the fastest of all the extreme clippers built for the transport of miners during the California Gold rush.

There's a list of other famous American clippers but with the American Civil War, the development and construction of clippers had moved to Britain. Using their industrial advantages, the British began building composite ships for the China tea trade. Planking secured to iron frames made them extremely strong. All riveted iron ships would be following in a decade. Smaller than the extreme Clippers, the Tea Clippers were crewed and sailed like racing yachts, and racers they were.

Extreme Clipper developed at the time of the California gold rush.

Tea Clippers---Canton to London in 100 days

Never had the world witnessed races of such a grand scale, races that spanned the globe, as those between the Tea Clippers. These races, sailed in upwards of a hundred days, were the Grand Prix's, the Derby's of their day and the merchant ships were sailed by large picked crews.

Captains and men alike who believed in piling on sail and driving their ships day and night through all weather. Among those who love the age of sail, the Tea Clippers names remain legend today: *Ariel*, *Thermopylae*, *Cutty Sark*, *Taeping*, *Serica*, *Fiery Cross*, and *Taitsng* just to name a few, and these ships were as closely matched as todays ocean racing yachts.

There was no official starting gun in the race between Canton, China and Atlantic ports like London, but races they truly were. It started with the loading of cargo, each ship loading and stowing thousands of tea chests as fast as possible. Hatches sealed, clearance in hand, each ship was out of port at the first possible moment bound for the far away North Atlantic. Profit and glory, prize money and bragging right and often ships sighted each other en-route, encounters that motivated crews to drive their ships even harder, at times arriving on the same tide after 18,000.

One of the best-remembered of all the Tea Clipper races was in 1866. Unfortunately, the Civil War had drawn the American clippers out of the competition, so the race was between the first five of the approximately 40 British clippers that sailed from Chinese ports that year – *Ariel*, *Taeping*, *Serica*, *Fiery Cross* and *Taitsing*. ARIEL, a composite clipper built of wood and iron with a sail area equal to nine tennis courts, was the favorite of the merchants. This favoritism translated into fast processing of her papers so that she left the dock at 5:00 p.m. on May 28. Unfortunately, the tug sent to escort her to sea was not powerful enough to clear the river bends with ship in tow. The *Ariel* was forced to spend a maddening nights wait for another tug.

Thermopile

The Captain of *Fiery Cross*, seeing *Ariel* depart, frantically drove the Chinese coolies to finish loading his cargo. Such was his hurry that he sailed without signing his bills of lading, but he did get a 12-hour lead over the rest of the fleet. Two of the others, *Serica* and *Taeping*, moved out almost together, leaving the harbor just after *Ariel* about 10:30 the next morning. *Taitsing* sailed on the morning after, followed by five others, who fell far behind during the three-month race.

The five leaders made good time across the Indian Ocean, *Ariel* logging 330 miles in one day and *Fiery Cross* a close second with 328 miles in a day. Every yard of canvas was up on every vessel, ready to make the most of the slightest breeze. *Fiery Cross* was the first to reach the Cape of Good Hope, with Ariel only two hours behind and *Taeping* 14 hours back. Into the Atlantic, they drew close together, although not close enough to sight, but all still straining forward with every timber and sail. At St. Helena the order was *Taeping, Fiery Cross, Serica, Ariel*, and *Taitsing*.

Passing Ascension Island, *Ariel* was catching up, passing *Serica* and crossing the equator at the same time as *Fiery Cross* and *Taeping* on August 4th. By that time *Serica* had dropped two days behind, and *Taitsing* was eight days back. North of the equator, *Ariel's* captain swung her westward on a hunch that there lay the best breezes.

Tea Clippers in company

Taeping and *Fiery Cross* remained on a straight course and within each other's sight for the next two weeks. At that point, they entered the doldrums and there, *Fiery Cross* lay dead in the water, completely becalmed for 20 hours, while maddeningly, though not far away, *Taeping* picked up a slight breeze and in five hours had disappeared over the horizon, leaving *Fiery Cross's* crew watching in helpless frustration.

By this time the sleek and beautiful *Ariel* was far ahead of *Fiery Cross* and *Taeping*. The *Serica* was now in fourth position and *Taitsing*, although still running last, had made up two days since the Equatorial crossing. On August 29th, the first four passed the Azores with *Ariel* still in the lead. By remarkable effort and good fortune, the Azores with *Ariel* still in the lead. By remarkable effort and good fortune, the *Taitsing* had now made up an additional three days' time.

Past the Azores, the wind was strong in the quarter. They ran to the shelf in five days, drawing closer together as they neared the Channel. *Ariel* was the first at Bishop's Rock, with *Taeping* close astern. Tearing up the Channel with all royals, stun-sails and kites set, they did 14 knots in the strong southwester. An hour ahead at dusk, *Ariel* ran on, burning blue flares as a warning, and arriving off the Downs before dawn, sent up rockets to signal the pilots. Taeping, closing and signaling

also, ran down on *Ariel*, now hove to for a pilot. Ariel bore up and ran between *Taeping* and the station, lest the second ship get the first pilot. Two cutters set out from the station, and *Ariel* ran in again, swung up and backed her jibs to board the pilot. Stepping aboard, the pilot whipped off his hat and saluted the *Ariel* as first ship of the season.

The captain, unable to repress his agitation, exclaimed "Yes, and what is that to windward?" He pointed to the *Taeping*, whose pilot was also boarding, and said, "We have no room to boast yet!"

No Room to Boast

Ahead by a mile as they breasted the Deal, the Ariel took in all sail with a tug ahead. Within the hour they discovered that *Taeping*. had picked up a faster tug. Emotions were feverish and tension high after 100 days of all-out racing and prize money close at hand. As the legend goes, the bosun and half the crew begged to crawl down the hawser to the tug in order to aid the stokers and sit on the safety valve!

Towing up the Thames

The race ended with *Taeping* as the winner by 20 minutes, due to a faster tug, not fair, a quirk of fate. The spectacular see-saw battle had the city in an uproar. It was the talk of London for days afterward. The owners of *Ariel* and *Taeping* saw no clear-cut victory in this great race, which all agreed should have ended at the Downs. They, therefore, decided to call it a tie and to divide the stakes equally between the valiant crews.

The Tea Trade

The tea which inspired such herculean sea dramas has been the center of a highly developed culture and ceremony in China since 800 A.D. Legend has it that the Emperor Shen Nung, in his travels once encamped beneath a camellia tree. Being hygienically. enlightened, he began boiling some drinking water. On this evening some of the camellia leaves fell into the boiling water and sent up a fragrance which tempted the Emperor. Liking the pungent flavor, he added more leaves from the tree, and tea was born – tea which William Cowper centuries later eulogized in:

"*Now stir the fire, and close the shutters fast,*

Let fall the curtains, wheel the sofa round

And while the bubbling and hissing urn

Throws up a steamy column, and the cups

That cheer but not inebriate, wait on each,

So, let us welcome peaceful evening in."

Tea reached the British palate by a circuitous route, the exotic leaves carried in small quantities by Jesuit missionaries. The effect was instantaneous. Queen Elizabeth immediately set about to contact the Chinese Emperor in the interest of direct trade relations. She dispatched three ships in 1596 under Captain Sir Robert Dudley, who carried a letter from his Queen to the Emperor of China. Unfortunately, this little fleet was never heard from again.

Undaunted, the Queen granted a monopoly to the Honorable East India Company on trade between the Cape of Good Hope and the Straits of Magellan, with hopes for the company's early success in dealing for China's tea. However, China remained implacably closed to commerce with the "foreign devils" for many years yet. Finally, in 1685 Emperor K'Ang Hsi relented, and England received its first shipment of tea directly from China.

The East India merchantmen took up to two years and more to make the round trip out to Canton for the tea. For this interminably long journey, the vessels carried livestock and fowl and even so had to stop en route for re-provisioning. They also carried important passengers for whom they provided luxury accommodations on the glamorous tea run. The East Indiamen, therefore, were heavy and roomy – impressive to look at but not designed for speed. They were built of timbers seasoned in saltwater until they were almost as hard as iron and fastened with massive copper bolts. As the pride of the East India Company, the vessels were kept dazzling clean and freshly painted and varnished. Sterns were elaborately ornamented; bows adorned with striking figureheads. Their officers' uniforms matched their beautiful ship in magnificence. Inspection officers wore white gloves, and woe befell the crew if the gloves were soiled in touching the brassware

East Indiaman Ship built for cargo capacity not speed.

This, then, was the entrenched and elite fleet that ruled the waves from 1600 until the mid-1800's when the clippers appeared.

On the morning of Dec. 3, 1850, London Town awoke to a marvelous sight. In the harbor lay a vessel so sleek and lovely it was like a great, graceful bird at rest. It was the American clipper *Oriental*, just 97 days out of Hong Kong with a cargo of 1,600 tons of tea. The record was astounding! The *Oriental* made the British East

Indiamen look like scows in comparison, with their sturdy, wide beams and blunt prows.

The newcomer had destroyed all British records for the tea run from the Orient, and she opened on that day the Tea Clipper era. There was admiration as well as consternation among the crowds of Englishmen appraising the lean, clean lines of the *Oriental*. The "London Times" pointed out with alarm that Britain must have faster vessels, or the Tea Trade would certainly fall to the Americans. British shipbuilders lost no time in designing vessels to match the American racers. Actually, the British had long known how to build for speed, but they had only built racing yachts of 100-200 tons instead of ships.

A few months after the British were stunned by the *Oriental's* challenge to their traditional East Indiamen, work was nearing completion on their own first clipper. This was the little *Stornway*, 506 tons, sent into service early in 1851, with the *Chrysolite* and others. following shortly thereafter. The Tea Clipper race was on, and the clippers regularly made and broke new records to and from the China Tea ports.

The glory rested not exclusively in the design of the ship. Much of the success was due to the spirit of the officers and crew who could push their vessel to its limit and hold it there for up to three months, repairing and maintaining it en route. They were bold men and brave, rough and ready and fanatically attached to their vessel.

The sailing time to China was not much different from New York or London because the shortest route was not the fastest. Frequently the British swung wide out in the Atlantic to gain speed. Going out, both nations sailed far south off the Cape of Good Hope to get into the "Roaring Forties", which sped them across the southern Indian Ocean, often with cargoes to Australia northward through the East Indies and into the China Sea. The success of the trip even the lives of the crew, depended on the master's skill, memory, instinct and raw courage in negotiating the rocky islands and treacherous currents of the China Sea which had not yet been precisely charted.

As the clippers arrived in China and received their cargoes, the crews raced to ready their ship for the return voyage. The clippers had to be in one of the trade ports by mid-June, but if they loaded too early, they would get part of the previous year's crop. Extra care was taken to load the tea because it is easily damaged by either air, water or fumes. Then their race for home began. Meanwhile, in England and America, there would be weeks of suspense, with excitement mounting and bets, prayers and boasts increasing daily. This, in the days before telegraph and radio there would be no word of the ships in the race until the final day, when stations on the shore spied the frontrunners and sent reports up the coast and inland. The news would spread like wildfire that the tea clippers were arriving. Some would hurry down to the coast to watch the ships as they picked up pilots and raced on toward the Thames. They were a magnificent sight, canvas wings trimmed to use every gust, every whisper of wind, clouds of sail speeding over the sea at the end of their homeward rush.

Once at the docks, the tea would be sampled; then bidding would begin on the thousands of chests of the delicious leaf were bargained over and bought. The generous cash prize awaited, provided for the winning crew because the cargo of the first ship brought the highest price for it was a social coup to offer guests tea from the season's first ship.

The story, the high adventure of the tea's journey and the romance of the racing ships seemed somehow to bring a thrill to teatime. The digging of the Suez Canal gave an enormous advantage to the steamships of the day. Not having to steam around Africa, gave steamers the economic edge that eventually took the Tea Clippers out of the Tea business.

The Raiders

Scourge of the shoreline

Late Minoan Galley

They could come rushing from the sea at any time in light fast vessels. Shoal draft and propelled by sail and oars, any river, creek or beach would afford a place to come ashore and plunder. Food, treasure and slave---slaughter and burn what you didn't take. The sea people are what historians call them now. Ancient seafaring races who lived by raiding along the shores of rivers and seas. They were the scourge of coastal communities, records of them exist before 2500 BC. Around 1800 BC the Minoan King of the island of Crete built a navy to subdue them in his region of the Mediterranean Sea. Without security villages, and commerce they couldn't prosper; a King couldn't ensure revenues. King Minos was successful until around 1650 BC, a volcanic eruption and tsunami destroyed the island of Thera and its harbor and fleet. The Mediterranean was swept by days of repeated tsunamis and buried with ash. The disaster halted Minoan trade over much of the Mediterranean and diminished the civilization and the social order.

In Egypt, there was inscribed on a clay tablet during the reign of Pharaoh Akhenaton, 1350 BCE. It was a report of sea raiders and nearly a hundred years after, on the *Tanis Stele*, Ramesses II, had recorded.   (*The unruly Sherden, (sea people,) whom no one had ever known how to combat, came boldly sailing in their boats, from the midst of the sea, none being able to withstand them.*)

Ramesses III must have had some success for on the north wall of the temple *Mediante Habu*, there was a base relief mural placed and inscribed with hieroglyphs, illustrating the sea people's arrival, raiding activities and his campaign to defeat them. Its known as the Battle of the Delta.

It is maintained that the Sea Peoples were an ever-shifting league of seafarers that gathered to raid not only Egypt but all the Mediterranean during the Bronze Age Collapse linked to the volcanic explosion. Their origins remain unidentified, but they seem to have been everywhere. Egyptian hieroglyphs next to prisoners translate only as Philistines, Tyrrhenians, Lycians, Pelasgians, Sardinians, Teucrians, Mycenae to name some. Anyone with a big boat? Well, we certainly have plenty of texts suggesting early Greeks were involved for they considered raiding an acceptable profession. Every man with a boat and a spear thought it was a great way of life.

Greek Galley like those described in the Iliad

"Along the ancient Aegean shores raiding was a perfectly normal occurrence, and a major source of income. No dishonor was connected to raiding, rather it was a route to the achievement of glory. Homer wrote it so in both the Iliad and Odyssey.

"We boldly landed on the hostile place, and sacked the city, and destroyed the race, their wives made captive, their possessions shared, and every man found a like reward."

Men thought the same behavior to be proper a thousand years later in 314 BC. As Alexander the Great conquered the lands bordering the Mediterranean, he discouraged raiding. When the captured captain, one of a sea raiding ship was brought before him, Alexander asked the man what right he had to raid and rob on the sea and its coasts?

The mariner answered boldly, *"What do thou meanest by seizing the whole earth? Because I do it with a petty ship, I am called a robber, whilst thou who dost it with a great fleet and are thus called emperor."*

Needless to say, Alexander had little success with eradicating raiders, nor did others that ruled after, including Imperial Rome.

Romans had to deal with the Illyrians, a nation bordering the eastern Adriatic Sea. The coastal Illyrian raiders had developed their own type of vessel, the *Lembus*. It was a light vessel, fast and built to be capable of navigating very shallow water, 3 feet. It could carry 50 fighting men in addition to the rowers. It could unexpectedly emerge from hidden inlets to make a surprise attack or retreat into shoal water to escape heavier vessels.

Illyrian Lembus

Rome's attack on Illyria pirates

The Romans fought two wars to eradicate them but although they conquered the country, they never enjoyed complete success with the pirates.

War and migration had a lot to do with the development of a different sort of boat in Northwestern Europe.

Throughout human history migration has been the norm rather than the exception and these mass treks of tribes and peoples are well recorded in the ancient world. For multiple reasons there has been wave after wave of migrations from the extreme Northwestern limits of Europe, nor is this pattern new. A genetic study focused on the island of Crete found that the ancient Minoans of 2500 BC shared the genes with the Neolithic inhabitants of Norway and Sweden.

Scandinavian countries do have severe weather and much of the soil is poor for agricultural. A society that in the best of times on the bare edge of survival can slip into starvation with a single bad season. In the Baltic Sea the Island of Gotland recorded this scenario in the *Gutsaga*, an oral history of when the island was no longer able to produce enough food to feed its population. The

Gotlander's king had them draw lots to see who would embark on a trek seeking new lands. According to the saga, a third of the island's population was selected and thus began their journey. Though some began on boats, most were overland journeys, often lasting decades. One of the earliest and best recorded was the Cimbrian Migration.

Cimbrian Migration

Toward the end of the second century BC there was a cooling of weather, combined with a rise in sea level that stressed the tribes inhabiting what is now Denmark and Northwestern Germany's, Jutland. Much of that area is low and swampy with sandy soil that produced barely enough food to feed the population in the best of times. Faced with bad weather and repeated flooding, a large portion of three tribes made the decision to trek south. We now know why. Analysis of human bones from those years show the people were starving in nine out of ten years.

The Roman historian Strabo wrote: *"And Poseidonius*, (God Poseidon) *also conjectures that migration of the Cimbrians and their kinfolk from their native land was due to a sudden inundation of the sea."*

Ephorus, also had something to say in his history: *"the Celti, though trained in the virtue of fearlessness, meekly abide the destruction of their homes by the tides and then rebuild them, and thus they suffer a greater loss of life as the result of water than of war."*

In 113 BC the Cimbri, Teutons, and Ambrones had had enough of Jutland. They began a migration south seeking land to sustain them. In was not a peaceful journey, for they were not a peaceful people. They fought, pillaged and took what the needed to survive on a twelve-year trek. Their route took them to the edge of the Roman Republic, at Noreia, near the border of present-day Austria and Hungary, they were met by Roman Legions. They sent a message of negotiation, the same offer was made at every meeting with Roman forces: the offer was always, *"give us land, for alliance, or meet our blades."*

At the first meeting at Noreia, the Romans laughed and attacked the ignorant barbarians. The legions were over-run mauled, swatted aside like children and only saved from complete annihilation by a severe lightning storm. Fortunately, the Cimbrians assumed the lightning was due to Odin and Thor being upset over something and broke off the battle. Not only were the northerners fierce, they were on average, eight inches taller and thirty percent heavier than the Roman soldiers. The Roman officers were in shock.

For the next ten years the Cimbrians trekked through what is now, Austria, Switzerland, Germany, Belgium, France, Spain, coming out victorious in every encounter with Roman Legions. In 101 BC they split in two columns to cross the Alps into Tuscany. The Romans with a reorganized army destroyed the Teutons at Arausio and then the Cimbrians at Aqua Sextiae. The men fought to the death and seeing their men defeated, the women killed their children and themselves rather than be taken as slaves. The Romans got little more out of victory than a sigh of relief, but this was only the start. The Celtic and Geomatic tribes north of the Rhine and Danube were always a problem.

There were many campaigns during the republic and the empire and holding territory north of the rivers never worked out. In early 6 AD a massive over 100,000 men, 13 legions plus civilian cargo transporters marched north and defeated King Maroboduus of the Germanic Marcomanni, and his alliance of tribes. General Varus was sent to whip the new province of Germania into order. As it happened, the province of Illyricum, (todays Austria, Slavonia and Croatia) revolted, with the aid of some of the Marcomanni. The Roman Emperor was forced into recognizing King Maroboduus, so he could disengage his legions to deal with the great rebellion in Illyricum. Varus now had only the 17th, 18th, and 18th Legions.

Varus was known all over the empire for his ruthlessness. He'd crucified over 2000 Jews while in Judea. His management style was the same and he made an overbearing governor for the new province. The German tribal leader Arminius, who was a Roman citizen and commander of a Roman auxiliary cavalry unit made up of Germans only pretended to be a loyal officer. Using the collective outrage over Varus' tyranny, disrespect and malicious brutality to the subjugated, Arminius slowly united disorganized tribes who had succumbed to hated Roman conquest. The secret alliance waited for the favorable moment to strike. Although cautioned against trusting Arminius, Varus ignored the warning and on a march to winter quarters, Varus and all three Legions were ambushed and destroyed in the dense Teutoberg Forest in 9 CE. The battle went on for three days and 20,000 Romans fell. Along the northern border see-saw actions went on for hundreds of years but in 9 A.D., after the defeat of Varus' legions, the Romans began building a defensive line, from North Sea to the Black Sea, stretching along the Rhine and Danube Rivers. It is known as the Limes.

River cargo barge common on Danube and Rhine in Roman era.

This border was practical in many ways. Logistically, legions on the Rhine could be supplied from the Mediterranean in the south via the Saône, Rhône and Mosel. This required only a short portage. Armies in the east could be supplied upriver from the Black Sea. The Rhine was already supported by the economy of towns and large villages before the time of the Julius Caesar's conquest of Gaul. The area paid for itself as the lands north of the line were undeveloped and a constant military problem. On shore there were fortified cities and camps. On thousands of miles of rivers, Rome maintained a fleet of patrol craft crewed by Roman Marines, or sometimes Romanized Tribal Auxiliaries, who convoyed cargo ships and made raids on the northern bank when it became necessary to deal with aggressive Germanic tribes.

Forty foot Limes river patrol boat manned with marines.

This system worked well for over 300 years. To control the rivers, special naval vessels were introduced. Roman river patrol boat manned by Marines. These boats equally effective for defense and raiding were from 40 to 60 ft long and manned by 20 to 50 soldiers. They could be sailed with and across the wind and were rowed at around 6 Knots. The ram on the bow also acted as a landing ramp when the bow was run on the riverbank. Boats, almost exactly the same, except for a raised bow were used by the Romans for dispatch across the channel to Britain. The tribes living on the rivers and coast of the North Sea and Baltic were building boats and they shared building methods.

Limes river patrol boat manned by Roman Marines, fast and manauverable.

Angles Saxons Friesians and Jutes

The Angles and Saxons were brought to England as Mercenaries as *foederati* or federate troops.

In 84 A.D. the Romans finally succeeded in subduing Britain after forty years of resistance and it became a Roman providence and the people became Roman citizens. Over the following three centuries, outside of Picts raiding down out of what is today's Scotland, and the occasional revolt, Britain remained a relatively peaceful place. In 128 BC Emperor Hadrian had a 60-mile wall built from the North Sea to the Irish sea to discourage the Picts from raiding. In 142 BC another wall, this one north of (Hadrian's Wall) was finished; it was the (Antonine Wall) which stretched roughly between modern Glasgow and Edinburgh, Scotland. The thirty-mile wall ran from the Firth of the River Forth, to the Firth of the River Clyde, separating the Scottish Highlands from the somewhat pacified Caledonian lowlands. It was also a shorter line to defend but after twelve years of construction it was abandoned 8 years later. It was never peaceful in the north and Hadrian's Wall remained manned by legionaries.

Small cargo vessel common on coasts and rivers

After the 180 AD death of Marcus Aurelius, the last of the five good Emperors, the empire began to slowly come apart. Disputes over who would be emperor, assassinations, Roman army against Roman army, weakened the empires ability to deal with invading barbarians. It took centuries but after the rule of Marcus Aurelius it was all downhill. Men to form armies were always in demand. The Roman Empire had been using troops hired from outside the empire for centuries. They were taken into the legions and became citizens but in later times irregular forces were contracted for wages, part of which might be food and equipment. The term for these units were **Foederati** or federate troops.

Due to revolts, civil conflicts, and tribal invasions and sporadic chaos, empirical authorities had a consistently increasing need for trained soldiers to protect the Roman heartland. In 383, Britain was loyal and far from real trouble. The danger was to Rome's richest and most populous territories in the eastern half of the Empire. In 383 AD the **Roman General, Magnus Maximus** revolted against Western Roman Emperor Gratian in Britain and was proclaimed Emperor by his legionnaires. To fight Emperor Gratian, he began stripping the Legions from northern and Western Britain and sent them to prepare for war in Gaul. He replaced them with Saxons from the mainland and Scoti Irish, all *Foederati* troops, who were stationed in the north and in Wales. He left the island defended by warlords.

Magnus Maximus sailed to Gaul (modern France) where he defeated Emperor Gratian. Emperor Theodosius proclaimed him the Western Roman Emperor, responsible for Briton, Gaul, and Spain. The ambitious Maximus marched on Italy two years later; he lost the war and his life.

Those Legions did not return to Britain.

Roman era small cargo boat, about 40 foot.

The Legion drawdown led to a frequency and intensity of incursions by Picts and Irish Scoti raiders. In 410 A D, Roman-British subjects, weary of attacks by increasingly bold barbarians, expelled the officials and wrote the new emperor, seeking aid. Emperor Honorius replied that he was in a desperate war with the Visigoths. He had no troops to send. They received help for a short time in the way of a special task force but when it was withdrawn, they were advised to muster their own defenses. Regrettably, the long peaceful locals, weren't able to unite or govern themselves. The task of organizing a public defense was completely beyond them. The barbarian raiders wreaking the greatest mayhem were the Scottish Picts and Irish Scoti. The strategy of using one group of violent barbarians to fight off other barbarians was decided on. A contract was the struck with Saxon chieftains, who sailed with their troops to Britain Once established, they loved their new land, and seeing their Briton employers as soft weaklings, who required other men to do their fighting, they developed the greatest distain for them.

The Saxons began by complaining that the Romano-Britons had skimped on the monthly supplies they had been promised, a fair complaint. To resolve the dispute, the Briton leader, Vortigern arranged a meeting between Briton nobles and the Saxons, led by two chieftains, the brothers Hengist and Horsa. It did not take long for the Saxons to feel insulted and they settled the matter by abruptly pulling out daggers and slaying the Briton nobles. Vortigern, being a friend of Horsa, was spared. The Saxon avowed that the Britons had made the agreement null and void by failing to honor its terms. Britons had made the agreement null and void by failing to honor its terms. Calling on their kin from across the North Sea they launched a seaborn invasion that engulfed Roman Britain coast to coast. Wave after wave of long ships landed on coast and sailed up rivers. Ultimately, Hengist and Horsa compelled Vortigern, now reduced to a mere puppet, to cede large regions of southeastern England to the Saxons.

Saxon Longship about 55 foot. 400 AD invasion of Britain.

40 foot Saxon longboat 350 AD
still carvel planked like Roman boats.

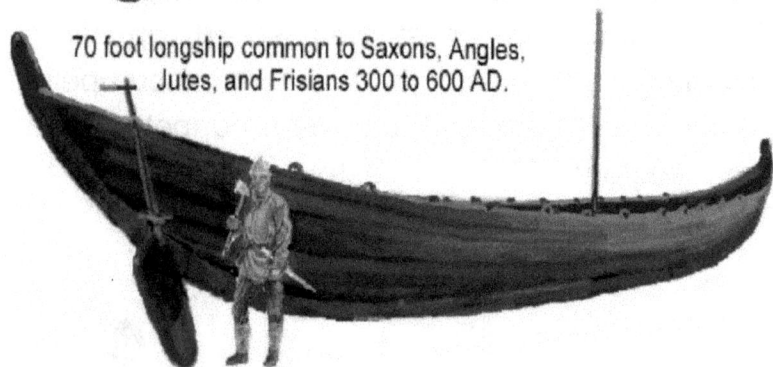

70 foot longship common to Saxons, Angles,
Jutes, and Frisians 300 to 600 AD.

50 foot Longship 500 to 700 A.D., the hull is riveted but
the bottom has a shallow curve. Oars on on rails with thule pins.

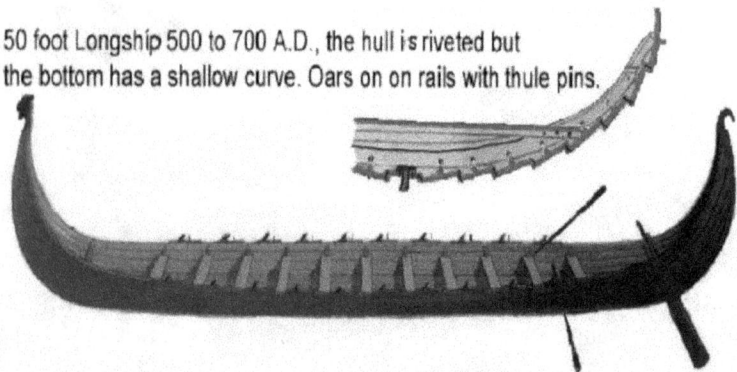

Of course, the Saxons and other invaders were not content and continued attacking other regions of the island to obtain land. They launched a war of conquest that sought to seize the entire island. They displaced and enslaved the local inhabitants, and replaced them with Germanic peoples like Angles, Friesians and Jutes, from today's Jutland in Denmark, and Lower Saxony in Germany.

 For Decades they just kept coming until finally there were 12 Saxon kingdoms, which the settled down to fighting with each other.

By then the last die-hard Roman-Britons had been driven into Cornwall and Wales where they doggedly held out till around 498 A.D. when they won a decisive victory at **the Battle of Mons Badonicus**. This battle and period gave rise to the legend of King Arthur, the Briton who rallied Britons against the Saxons. The results of that battle were the stabilization of the border between the Britons and Saxons, and their allied Angles and Jutes. For half a century afterwards, the Britons maintained the region west of a curved line running from Dorset on the English Channel to the Derwent River in Yorkshire and also included a few areas that protruding from the line near London.

Some form of Christianity was maintained in Britain after the Romans but was much more common in Ireland. In the sixth century it began to take hold among and the Saxons, and by 650 A.D. was widespread. For the next century the islanders mainly fought amongst themselves and with the Irish. For a change though, they had begun praying to the same God. As Europe went through continuous upheaval and the Roman Empire disintegrated, Britain was left mostly to itself until 793 AD.

The Viking Raiders

In northern Europe, it was different and yet the same. Since before the 1st Century, we know there were seafaring Germanic peoples both in Frisia, and all around the North Sea, in the Baltic. Large groups of raiders and settlers were entering the British Isles as the Romans Armies pulled out. We know from the sagas that every chief or landowner required his carls to build and maintain boats for travel, trade and of course, war or raiding. Small but seaworthy vessels, oars and sail were in use from the Baltic to the Irish sea.

They were similar whether built by Celts, Nordic peoples, Frisians, Saxons, Danes, Swedes, Finns,

Norwegians and more. By 793 CE, the date historians pronounce as the official beginning of the Viking period, the Norse had already been raiding into northern Europe, Asia, and the British Isles for a long, long time. Because the Nordic regions are bordered by the Atlantic, Baltic and North Seas, communication, trade, and travel were near impossible. No-one could travel far without having their advance frustrated by mountains, lakes, rivers, straits, a multitude of fjords, and the open ocean. Geography naturally forced the Scandinavians to develop a seafaring culture. Burial mounds with medals and large war canoes are dated back over 5000 years. Boats of up to 60' have been carbon dated around 100 CE and these were still planked with the edges butted. Around 250 CE

Nordic building techniques made a great leap forward. Lapstrake (or clinker-built) construction was developed, in which the planks of the hull overlap, progressing like steps, light but strong and flexible.

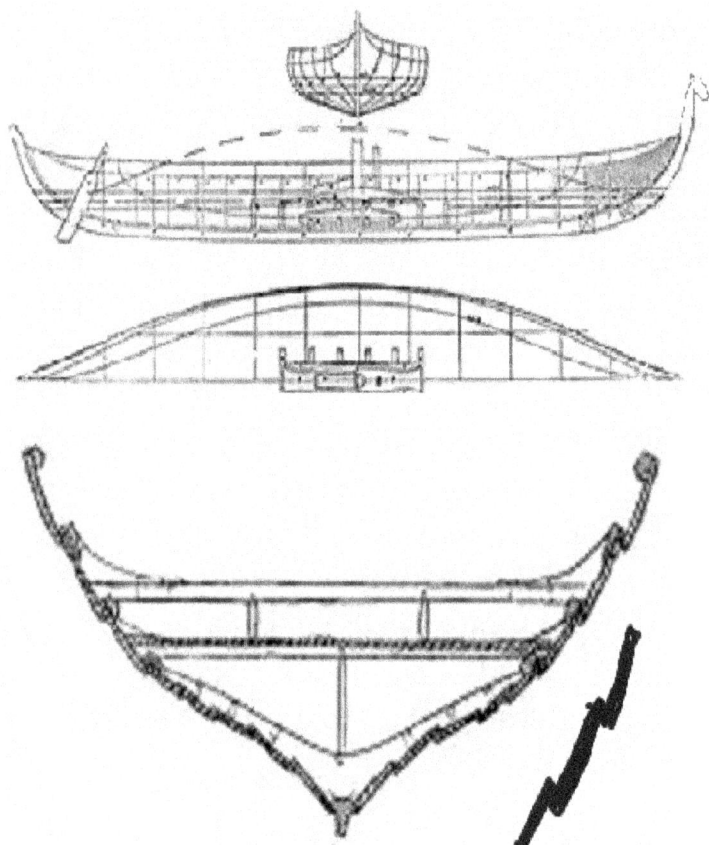

Lapstrake

Norse Long Ship 850 CE. Plugs for oar ports.

The boats were becoming larger, held together with iron rivets rather than being lashed and hulls were being driven by oars and sail rather than paddles. These innovations were refined and by the sixth century CE provided the sea-going vehicle needed to launch the Viking age. In this era Swedish Norse trading expeditions began to explore the waterways of what would become,

Russia, Ukraine, and the lands around the Black and Caspian Sea.

Longships Sailing Out of Fiord

Begining in the early sixth century, kingdoms rose in the old Uppsala area, the central lake region of Sweden, beginning what's known as the Vendel Period. The great clan leaders had wealth, raised well armored companies of elite mounted warriors, brotherhoods named for animals, the bear, wolf, hawk, etc. Their armor, bronze and iron were gilded and gleamed with settings of gemstones. The Goths that overran the Roman empire were an earlier migration of these peoples. In the sagas and Roman writings of the times, it is recorded that they had the finest horses and fastest ships.

During the Vendel period the Norse trading city of Novgorod was established first as a trade station between the Norse (called the Vangarians by the Greeks) and the Mideastern markets in the sixth century. Novgorod is considered the beginning of the Russian nation. It became a center for the trade of the iron, fur, ivory, amber, and of course slaves, all of which were traded via the Volga River and Don, to both the Black and Caspian Sea, 5,000 miles to Persia and the Muslims buyers.

Norwegians Swedes and Danes established other routes to the south entering from the Baltic at Riga to portage to the Dnieper River through Belarus and through Poland and portage to the Dniester, both routes down to the Black Sea and Constantinople to trade with the Roman Greeks. They established the great trading city of Kiev and raided everywhere they went. Seldom did a Norseman pass up a chance to convert the locals into profitable slaves. In return they brought silver, steel, silk, spices, fabrics, wines and many other goods back from the south.

The **Vendel** period merged with the Viking era around 800 CE marked by Christians recording horrific raids on isolated targets. Earlier raids are recorded but the Viking raid on the holy isle of Lindisfarne, on the Northumbrian coast was unique in Christian perception because it was an assault on the sanctified heart of the Saxon Northumbrian Kingdom. It was a desecration of the place where the Christian religion had begun in that land. Cuthbert had been bishop there and was revered as a saint; his body lay there. It was the most sacred place in all Britain. Of course, they surmised, it was no chance thing. God must be punishing them for a great sin.

They concluded that the sin was to be laid on the clergy because of a nobleman named Sicga, who with his henchmen, had murdered King Ælfwald of Northumbria and later who died by his own hand. His remains were carried to Lindisfarne and buried, thus a man guilty of both regicide and suicide had been interned on holy ground. Of course, Viking raiders had descended on Lindisfarne within weeks to deliver Gods holy vengeance and even further suffering was in store for his sinful people.

Reasons aside, the Lindisfarne raid---it was only the beginning of suffering, for this new wave of Nordic raids increased all around the coast of

Britain, Ireland and France. The Vikings preferred to attack along the coast and up rivers because these regions were impossible to protect. They would target monasteries and churches because there was usually gold. They would strip a place of booty and burn the rest, tactics which caused mass fear amongst the clergy and their flocks.

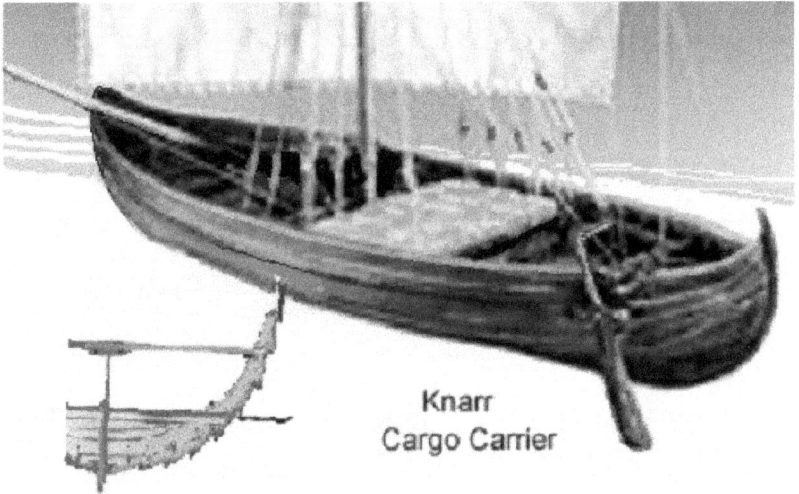

Knarr
Cargo Carrier

Vikings' success was for the most part due to the superiority of their shipbuilding. Their warships were made for speed and seaworthiness. There were three categories of warships or long-ships: *Dreker, Skeid* and *Snekke*. Built for Cargo and trading, the **Knarr** was wider, deeper and more stable than the narrow shallow warships. Meant for voyages on open oceans, the *Knarr* could be rowed in or out of port but were mostly dependent on sail.

A type opposite to the Knarr was the **Byrding.** The warriors and merchants who sailed to Eastern Europe had different needs in a vessel. The smaller, shoal draft, and very light Byrding was developed for the for the trading expeditions that used the rivers of northern

Byrding Longboat

Europe to reach the Black and Caspian Seas. The *Byrding* had to be light enough to portage, (be dragged overland between river systems) and navigate river shoals and rapids as well seaworthy enough to cross the Black Sea. Scandinavian ships were mainly built from strong oak, all with split logs, hewn planks that maintained the unbroken line of the grain, and resulted in light, flexible and very strong planks. The masts were kept short so as to allow rapid rigging and unrigging. With a low mast the longboats could also pass beneath most river bridges of the time.

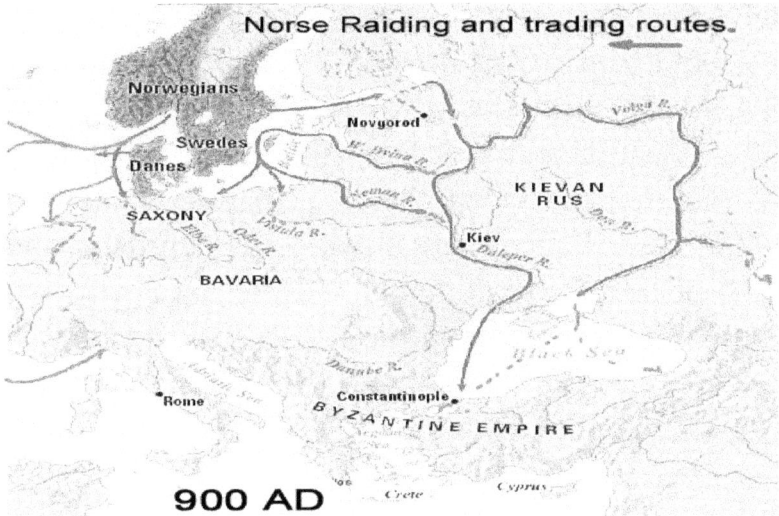

Norse Raiding and trading routes.

900 AD

Generally, about 60 feet in length, these long narrow ships could seat 40–60 oarsmen, as well as additional fighters. They could carry sizeable forces at speed and land wherever it proved advantageous. Because they were light and shallow draft, long-ships were landed directly on sand beaches or run into shallow rivers, avoiding fortified harbors. They could land practically anywhere, navigate rivers and portage between them. There were few places in the medieval world they couldn't go.

The design for speed was essential to their hit-and-run raids. As an example, when a large Viking fleet engaged in the sack of Frisia (now Holland and Belgium), Charlemagne's army marched on the region in haste to find it looted and burned. The Vikings were long gone and reportedly, already assaulting another region. Their ships are what gave Vikings an element of surprise, and also a way to disengage quickly. Travelling in small bands, and great fleets, they could remain undetected, striking and departing before reinforcements could even be called for. Their warships were mobility.

Political decisions between Norse leaders were usually based in law but also honor and wealth. Financial gain and practicality could be large factor. If two equal fleets or armies faced each other they might parley and decide on a neutralizing solution. It was far more profitable to join forces and raid to the west than fight with each other for their regional leaders. More than a few times enemy fleets, recognizing that they were near equal in strength, parleyed and agreed to raid England or France rather than fight each other.

Originally, the Vikings focused their attacks to "hit-and-run" raids along the coast and up rivers because these regions were impossible to protect.

By 814 CE sailing up the Seine and Lorie river, Vikings repeatedly looted towns and monasteries in Northwestern France and the Bay of Biscay. Within 60 years Norse armies were wintering in England instead of sailing home, and by 870 CE the Danes had conquered the northern, middle and eastern Anglo-Saxon kingdoms, Ivar Ragnar's son had Ireland, and Rollo's Norsemen controlled much of western France. At this time the Norse were also raiding and trading in the Mediterranean where they clashed with Muslim Saracens.

In the period between 600 CE and 750 CE the forces of Islam had swarmed from the Arabian deserts and subjugated Middle East, pushing east all the way to the borders of

Tibet, the Philippian Sea and west across North Africa, before crossing the Straits of Gibraltar to vanquish the kingdoms of Spain. Only the Franks under Charles Martel stopped them at the battle of Tours in 733 CE. Blocked by the Franks in the west and Roman Constantinople, with its navy in the East, Muslims armies were unable to penetrate into Europe and it remained Christian. To the south, the Mediterranean Sea was another matter. As the Islamic nations expanded their fleets, the Mediterranean became largely a Muslim Lake.

This was possible because the Roman empire of the west had officially collapsed in 476 CE. Beset for 300 years, by waves of barbarian tribes and internal wars, the west was in chaos. Charles the Great had formed a new Western Roman Empire and in 800 CE was named Emperor. The major focus of this new empire's rulers was inland, dealing with civil wars, and invasions all over central and western Europe; Viking raiders created some of these problems. The coast of the western Mediterranean was as in other eras, left to fend for itself

For a hundred years, opposed only in the Aegean and Adriatic by Roman Constantinople, and by its vassal navies like the Venetians, Muslim fleets had carried armies across the waters. They conquered the islands of Sicily, Crete, finally Malta and even large swaths of the southern Italian mainland. The independent Saracen raiders that sailed out from their strongholds were even a worse scourge. They used galleys similar to early Greek Penteconter, fast rowing vessels, around eighty feet in length and able to enter shallow rivers or to be beached if need be. The first peak of their strength was in the middle of the ninth century, fleets of Saracen slavers raided far and wide with little danger of Christian reprisal. So ravaged were some regions that vast stretches of the coasts were emptied and left uninhabited.

Viking raiders, aware of this confusion recognized opportunity and entering the Mediterranean at Gibraltar, sacked Muslim and Christian communities with equal fervor. A Viking fleet led by Bjorn Ironside, mistook Luni for Rome, and sacked that city, plundering its churches of their gold. They then sailed 60 miles to the Arno river and Sacked Pisa before venturing further inland for loot. On the way home they hit Muslim settlements on Sicily and the North African.

Other raids took place over decades, even raids made by Vikings sailing out of the Black Sea, still, the Mediterranean was for the most part a Saracens hunting ground until 999 CE. The first Christian Normans arrived as returning pilgrims from the Holy Land. Seeing weakness, they returned from Normandy in force. Soon they had conquered all of southern Italy, right to the border of the Western Roman Empire and under Roger Bosso, they also seized the island of Sicily from the Muslims. Christian rulers were regaining some control on land while Constantinople and maritime republics like Venice and Genoa were re-establishing some control over the Mediterranean Sea.

The three centuries of the Viking Era proper, characterized by widespread raiding and exploration, is a history written by their victims.

By 800 A.D. the Longship was all riveted lapstrake construction. The ships were stronger, longer, more stable, more flexible and far more seaworthy . They were able to make long ocean passages, land on beaches, run up rivers, and make portages.

oar port

Shallow V hull for speed and shoal draft, allowed beaching

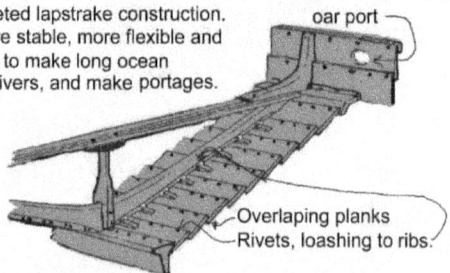

Overlaping planks
Rivets, loashing to ribs.

Eventually, the Vikings settled in the areas they had raided and overrun. They'd turned to farming and peaceful trade. The Viking era officially ended in 1066 at the battle of Stanford Bridge when an arrow pierced the unarmored chest of the greatest Viking of the time. **King** of Harald Hardrada. He was the victor of over 240 major battles from Persia and North Africa to the North Sea. Already king of Norway, Denmark and Sweden, Harald was also the true heir to the throne of England. He had a far better claim to the throne than William of Normandy or Harold the Saxon. The weakened Saxon army was then defeated by the Normans at Hasting, the Normans themselves being the descendants of Viking raiders who settled in France.

Saracen Raiders

Ottoman Saracen Galleys raided on the coasts of the Black Sea and up the rivers of eastern Europe.

The Mediterranean Sea remained in a see-saw conflict for the next three hundred years, naval dominance shifting back and forth between Christian and Muslim coalitions. In some cases,

Muslims and Christians allied with each other against mutual competitors. This changed in 1453 with the Fall of Constantinople. Within fifty years, Muslim fleets had captured the islands of the Aegean, the Adriatic and the countries of the Balkan peninsula. They controlled much of the Black Sea and had bases in Crimea, supported by their vassal, the Muslim Tarter Khanate. These conquests opened new floodgates for the slave-trade from in eastern Europe to the Caspian Sea.

In the Mediterranean as a whole, Ottomans supremacy was firmly established in 1532 after the pirate raider, Hayreddin Barbarossa, was named Grand Admiral of the Ottoman fleet. He and his brothers had for decades ravaged the Med, putting islands, cities, and great swaths of coastline under Ottoman control. A thousand books could not describe all the destruction and human suffering. When Barbarossa became Admiral, the conquests became methodical. An alliance and trade treaty arranged between the French King and the Ottoman Sultan lasted 80 years. In support of the

French King, Barbarossa devastated France's enemies and insured the Ottomans dominance of the Mediterranean until the battle of Lepanto in 1571. In the East, working in conjunction with the Tartars, the Ottomans raided the coasts of the

Black Sea and up the great rivers of Eastern Europe, capturing and more often purchasing huge numbers of slaves from the Muslim Tatar Khanate.

These slaves: Ukrainians, Poles, Georgians, Circassians, Greeks, Armenians, Bulgarians, Slavs, southern Russians, Greeks, Romanians, Transylvanians and Hungarians were **sold throughout the Ottoman Empire**. While records are not complete, the Ottomans recorded that between 1460 and 1690, tens of millions of men, women and children were taken in raids and military actions in Eastern Europe alone. Most were killed. Of those captured for the purpose of slavery, only a portion survived to reach market.

According to customs records, millions were sold within the walls of Constantinople alone. Even defeated at the Battle of Vienna, in 1683, the Muslim army returned with nearly a hundred thousand new European slaves.

Almost all slaves were sold into various regions of the Ottoman Empire, boys castrated as eunuchs, (not a good survival rate) women went to harems, men to various labors or to the galleys. The slave-raids occurred on an annual basis and reading 15th century accounts of them is disturbing. These raids are virtually unknown among Westerners.

They are seldom mentioned in the curriculum of educational institutions but are recognized by historians as the reason for Russia's repressed cultural and economic development in the 14th, 15th and 16th century. The Russians and Ukrainians are very aware of their history during this era. Along with the earlier Mongol invasions, the harvesting of the steppes prevented the expansion of settlements and populating of Russia and the Ukrainian steppes.

An account by S. Herberstein, the Ambassador of Emperor Charles V, is translated.

"He took with him from Muscovy in one raid, so great a multitude of captives as would scarcely be considered credible; they say the number

exceeded eight hundred thousand, part of whom he sold Kaffa to the Turks, and part he slew. The old and infirmed, who will not fetch much at a sale, are given up to the youths, either to be stoned, or to be thrown into the sea, or to be killed by any sort of death they might please."

Mikhalon wrote 1555,

"The Crimean Tatars have many more slaves than livestock and therefore they supply them also to other lands. Many ships loaded with weapons, garments and horses came to them, each after the other from beyond the Pontus and from Asia and left always filled with slaves. … So, these plunderers always are in possession not only of slaves for trade with other people but also have slaves for their own use and to satisfy their cruelty and waywardness. In fact, we often find among these unfortunate slaves very strong men, who, if not castrated, are branded on the forehead or on the cheek, and are tortured and tormented by day at work and by night in the dungeons.

The North African Raiders

Four brothers from the Aegean vastly altered the political map of North Africa. Beginning as traders, they soon became privateers and then outright raiders. Born on the Aegean Island of Lesbos, the brothers: Oruç, Khizr, Ishak and İlyas shook the world. Two died in fighting but the eldest, Oruc captured islands and the cities of North Africa for the Ottoman Empire. His younger brother, and lieutenant Khizr, known as (Hayreddin Barbarossa) went further after Oruc's death, taking on and defeating the fleets of the European states. He so cleared the seas of defenders that Saracen raiders could operate at will.

Though victims were not so plentiful as in Eastern Europe, the economy of North Africa's Barbary Coast, was built on raiding for and marketing slaves. The slave trade had been a pillar of the economy in North Africa even before the times of the ancient Phoenicians, with white slaves taken in European conflicts and black slaves arriving through trans-Saharan routes. The towns on the North African coast continued to be famous during Roman times for their slave markets. There was no reason for this to change in the medieval age.

The Barbary Coast increased in influence in the 15th century, when the Ottoman Empire took over from Christian rulers of the area. The Ottoman provinces of Tripoli, Algeria, Tunisia, and the Sultanate of Morocco, raided on land and sea up until 1830. European slaves were acquired by Barbary, raids on ships and by raids on coastal towns. Not only was the eastern Mediterranean under constant threat but from Italy to the Netherlands, Ireland and the Southwest of Britain, as far north as Iceland. Raids were large and small often landing on unguarded beaches, sneaking up on towns in the gloom of night to enslave their victims. Here's just a few examples.

Iceland. In 1627 two groups of raiders sailed from the Barbery coast, 3000 miles to Iceland. One group landed first at Grindavik, a small fishing where they captured no more than fifteen men. From there they sailed to the Governors town of Bessasaoir, where they were met by canon fire and having lost the advantage of surprise sailed for home with the few slaves they had.

The second group fared better. These raiders came ashore at Eastfjord, on the southeastern coast, the village Hvalnes, taking 115 slaves, gold and silver coin and other valuables. Encountering headwinds, they sailed along the south coast capturing two ships en-route. On July 16 they reached the island group of Vestmannaeyjar where they raided for several days. They enslaved 240 in town and more from the other islands, and killed anyone that resisted as well as those old, infirm, or lacking in market value. The Icelanders were mostly sold on arrival in Algiers. Only a few ever returned, these ransomed years later by the King of Denmark.

Ireland. Four years later, almost to the day, in the pre-dawn hours of June 20, 1631 Saracens from Algiers, ran their boats onto a beach of the village of Baltimore, on the SW coast of Ireland. They spread out through the village and on signal, smashed their way into every home, pulling the people from sleep into a waking nightmare. Men women and children of a useful age were dragged aboard the ships, the remainder of the population disposed of with sword or cudgel. After the voyage back to Algiers, the Irish captives of Baltimore were taken to slave market and postured naked before potential purchasers. Women were usually sold into harems or bordellos, men for hard labor. The children were often raised as Muslims and placed in the slave units of the Ottoman army. Baltimore remained uninhabited for decades.

England's coastlines were closer and more venerable to attacks from the Barbary raiders. England's south and west coasts were attacked at will. In the early 1600s the situation became so bad that between 1609 and 1616, England lost 466 merchant ships. Small fishing boats out of small ports were stripped of crew and left to drift.

In 1625 they raided Mount's Bay, Cornwall, the boats sweeping onto the beaches, taking 65 women and children. A year later St Keverne was attacked several times, and a portion of its population enslaved. Attacks along hundreds of miles of coast were daily occurrences. Another Barbary raid on the Cornish coast took 240 men, women and children, including family members of parliament.

In 1640, parliament set up and funded a commission to oversee the ransoming of captives. It was reported at that time, some 5,000 English people were enslaved in Algiers alone. Parliament sent Edmund Cason to Algiers to negotiate the ransom of English slaves. He agreed to a price of 3£0 per man, more for women who were naturally more expensive to ransom, but he only had enough funds for the repurchase of 250 of the enslaved English. While in negotiations, hundreds more English subjects were taken in raids.

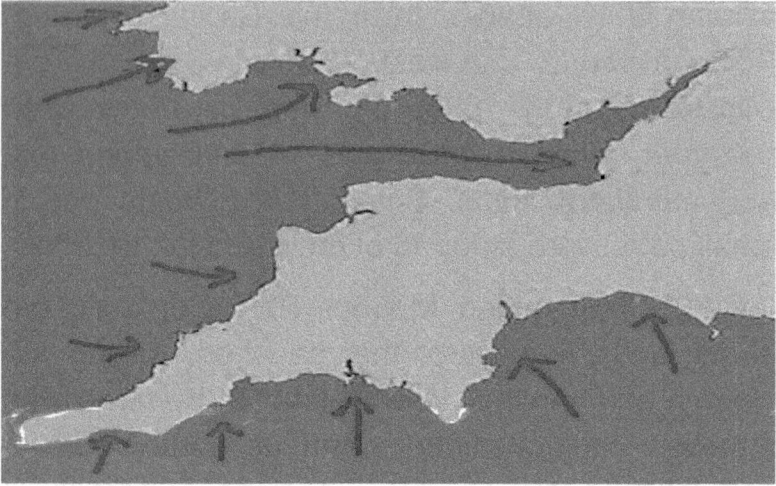

Barbery Slaving Raids, Cornwall and Wales

By the 1650s the attacks were so frequent that they threatened England's fishing industry with fishermen reluctant to put to sea, leaving their families unprotected ashore. When Oliver Cromwell dismissed parliament in 1653, he took personal action. He set up a naval force to deal with the raiders and decreed that any captured raider was to be brought to Bristol and in public, slowly drowned. Going further, he sent the fleet to the mouth of the Bou Regreg river, in the Republic of Salé, Morocco. The orders were that the port was to be bombarded, and it was but on the coasts of Dorset, Devon and Cornwall slaving raids continued.

During the same period, raids in the Mediterranean were so regular they had overwhelmed man's ability to populate the land. The coastline from Venice to Malaga suffered widespread depopulation, and coastal settlement was discouraged. In 1663 the Saracen raider "Turgut Reis," plundered the coastal villages of Granada, Spain enslaving 4,000 people. Off the northern coast of Sicily, the 9,000 inhabitants of Lipari Island, were captured and transported to the Barbary Coast, for sale. In fact, it was said that "there was no longer anyone left to capture. It is estimated that the slavers of the three cities of Tunis, Algiers, and Tripoli alone enslaved 1.25 million Europeans between 1600 and 1820. Why so many?

Depending on their owners, slaves could have short lifespans. Some galley owners sent slaves ashore to labor when ships were in port, others kept them chained in place to row until they died, eating sleeping and defecating on the bench where they rowed. Enslaved field workers, laborers of every sort, died faster than non-slaves. Only 10% of the boy castrated for service in Harams survived. It is estimated that an average loss of 25% a year had to be replaced. Plus, a large percentage died during the process of capture and transport. **90% of black Sub-**

Saharan slaves died crossing the Sahara Desert to Mediterranean markets.

Decline of the galley

Long after Europeans had retired oar propelled vessels in preference of heavy sailing ships, vessels carrying numerous long-range cannon, most Barbary ships remained galleys. Rowed by slaves and crewed by a hundred or more fighting men, armed with swords and small arms, they were most dangerous in a calm. The Barbary navies were not battle fleets but entrepreneurs. If a European warship appeared on the horizon, they avoided it.

The Barbary Chebeck

By the 1700s though, the Barbary ships had evolved from light galleys to more substantial vessels. They were now primarily sailing ships up to a hundred fifty feet, the size of European corvettes or British sloops of war. Most remained lateen rigged and though they had oars, they were merely oar assisted. The Chebeck, as it came to be known, was one of the fastest and best sailors in the Mediterranean, on a par with the American Baltimore Clipper type, although not as robust. Eventually most navies developed their own version of the Chebeck. The first knock-off was the Spanish navy, who built their Chebecks to fight the North African Raiders with their own design.

Chebeck under Sail 17th thru 18th centuries.

By the mid-1700s, the scope of piratical activity began to diminish as the more powerful European navies began to coerce the Barbary States to make peace and cease attacking their shipping trade. Merchant shipping and coastlines belonging to small Christian states lacked this protection and continued to suffer. Many chose to pay tribute.

With independence from Britain in 1783, the new Confederation of American States was no longer under the protection of the British Navy. American ships began to be taken by raiders from the Barbary States. Congress had **disbanded** the Continental **Navy** and the last ship; the Frigate alliance was sold in 1785. There was no navy to protect the merchant ships.

In 1785, Thomas Jefferson and John Adams were sent to London for negotiations with Tripoli's Ambassador. He was asked what right his countrymen had to rob and enslave Americans. The Muslim envoy explained that the "right" was "founded on the *Laws of the Prophet*, that it was written in the Koran that all nations who have not answered to the authority of Muhamad were sinners; it was their right and duty to make war upon them wherever they could be found, and to make slaves of all that could be taken as prisoners. Every Mussulman who should be slain in battle was sure to go to Paradise". Roughly 500 Americans were held captive in this region as slaves in 1785 and without a Respectable Navy there was dam'd little Congress could do about it.

The governing authority under the Articles of Confederation, was too weak to maintain more than a token armed land force of 700 men. American nationalists saw the necessity of adopting a new constitution that would increase the authority of the national government, principally by giving it the power of taxation. The need for naval power was accepted and needed little debate at the Constitutional Convention in 1787. The frame of government proposed gave Congress power to raise money to "provide and maintain a navy," but they took no action.

Both France and Britain were interfering with American shipping and the Barbary states were still raiding. In 1793, President Washington spoke to Congress of the nation's need to prepare to defend itself. Inside of a week, news reached Philadelphia of the truce between Portugal and Algiers, opening the way for Barbary raiders to cruise the Atlantic and imperil American trade. The obvious obligation of defending the nation's seaborne merchants finally motivated Congress to establish a navy.

On January 2, 1794, a naval force adequate to the protection of the commerce of the United States, against the Algerian raiders was ordered built. After three months of haggling the act passed the Senate, and on March 27th was signed by the

President. Six state of the art Frigates were ordered, and these vessels were to combine such qualities of strength, durability, swiftness of sailing, and force, as to render them superior, to frigates belonging to any European Power. The American leaders intended that the navy have superb ships of war.

USN Frigate Constitution

Three frigates were complete in 1797 when negotiations that agreed to annual tribute produced temporary peace. This deal only lasted four years, the tribute amounting to 25% of the Federal budget was unacceptable to President Jefferson, who in 1801 joined Sweden in what is known as the First Barbary War that ended in 1805.

The Second Barbary War (1815), was again fought between the United States and the Ottoman Empire's North African regencies of Tripoli, Tunis and Algeria. During the War of 1812 American merchant ships avoided the Mediterranean so as not to be captured by Britain's Royal Navy. But problems arose again with the Barbary States at the war's end in 1815. Concluding that the Americans had been weakened by war, the Dey of Algiers declared war on the United States and turned his raiders loose on American interests. President Madison informed Congress and immediately placed a naval squadron of ten ships under Commodore Stephen Decatur and William Bainbridge, both veterans of the earlier Barbary war.

By July 1815 Decatur's ships had captured several Algerian ships and forced the Dey of Algiers and Tunis to release all American and European slaves and commit to treaties. Attacks on American merchant ships were effectively ended. Additional problems with the United Kingdom and the Netherlands marked the beginning of the end of raiding in that region. France conquered the North African states 15 years later. A thousand years of organized raiding had come to an end, at least in North Africa.

The American Whale Boat Raiders

Summer of 1776, it was the beginning of the American Revolution. At the southwestern end of Long Island, a British fleet landed an army of 32,000 men. The outnumbered 10,000-man Continental army was defeated at the battle of Guan Heights. The defeated Continental forces retreated to a defensive position on Brooklyn Heights and the British

Thirty foot Armed Whale boat

The New Jersey whaleboat men who braved the might of the Royal Navy to carry the war to the enemy.

began preparing for a siege. On the night of August 29, using the ruse of campfires and music,

Washington evacuated the entire army to Manhattan. He did it without the loss of supplies or a single life. Further defeats forced a retreat through New Jersey to Pennsylvania. For the rest of the Revolutionary War, the British held Long Island.

General Richard Howe and General William Howe organized the southern tip of Long Island and Manhattan Island, (now the New York City Metropolitan area) into the center of the British political and military operations in North America. The harbor thronged with military activity as countless British and loyalist watercraft carried farm products from Loyalist farms on Staten Island, Long Island and New Jersey, navigating the Kills, the lower Hudson and Harlem rivers. Men-of-war patrolled and convoyed, acted as guardships protecting the merchant vessels that were supplying the British Army now isolated in New York. The Americans, lacking a navy and the funds to build one, debated the issue in Congress. On April 3, 1776, Congress passed a bill, for the issuance of **Letters of Marque**.

"INSTRUCTIONS to the COMMANDERS of Private Ships or vessels of War, which shall have Commissions of Letters of Marque and Reprisal, authorizing them to make Captures of British Vessels and Cargoes for profit."

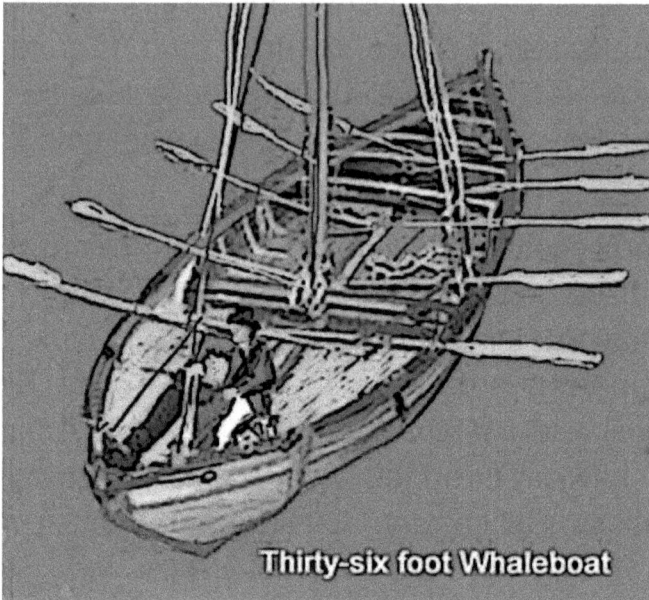

Thirty-six foot Whaleboat

Immediately vessels large and small on the American coast began to bristle with guns. In Long Island Sound and on the Jersey Shore many large vessels were converted or built as privateers but the whaleboat was the local vessel of choice. Watermen began to build enlarged whaleboats, 30 to even 40 foot in length, light, double ended lapstrake boats, propelled by oars and sail that could function in any coastal area and in all weather. These were boats that any Viking raider would have been at home aboard. With no more than a little swivel cannon (*Glorified Shotgun*) mounted forward, they moved fast and silent. Crewed by 12 to 30 men, the core of which would be local watermen with intimate knowledge of the region, the whaleboat men armed themselves with muskets, boarding pikes, pistols and cutlasses, but stealth was their greatest weapon.

The population of New Jersey and long Island was by no means united in its support of the rebellion. Loyalists seized property and crops from rebels and transferred it to British ships for transport to New York. Rebel raiders stole it back and supplied the Continental Army. The coasts of New Jersey and long Island sound provided perfect locations for small American privateers to exasperate and attack the might of the British Navy. Patriots used their access to the myriad bays, rivers and ocean inlets to disrupt British shipping, and to attack both British, and loyalist vessels. On landfall off New York, British merchant ships would stand off the coast near Sandy Hook or anchor off Staten Island, under the cover of shore guns and ships of the Royal Navy.

Often, under the guns of big ships and forts the whaleboat raiders made swift attacks undercover of dark, often in rain or snow. They would position a number of boats and on a signal act in concert, like swarming ants. They would load and make their escape, navigating hazardous bars, and shallow inlets, disappearing up twisting river channels where the well-armed and deep draft warships were unable to pursue.

Raids in Long Island Sound were so successful that soon Long Island Tories built their own whaleboats and began raiding on the Connecticut coast. Whaleboat raiding by both Patriots and Loyalists across Long Island Sound began in 1776 and was still going on in 1782. The initial plan was to patrol in order to prevent smuggling between the Long Island and Connecticut shore but soon a far more aggressive policy took over. Crews began to land on Long Island and loot any and all British property within reach. Of course, it was only to be British or Loyalist property depending on which side was doing the looting and on return, all captured goods were required to be listed and reported for shares and taxes. As you may have guessed, this seldom worked in practice.

A letter from the Connecticut Governor, to Whaleboat commanders read, (*"Whereas sundry and repeated complaints have been made that persons under authority of commissions given to armed boats to go on shore on Long Island to act against the enemy there have unjustly and cruelly plundered many of the friendly inhabitants there, brought off their effects, and have not* *caused them to be listed and condemned in course of law).*

There were so many complaints that the State Assembly appointed a committee to sort them out, but this proved difficult. Anybody with a boat and a few dishonest associates could imitate a Viking raid across the sound and most knew enough to keep their mouths shut about it afterward.

An officer of the Continental Forces, working as a spy on Long Island complained that the whaleboat raids had gotten completely out of hand. He said that if the weather was good enough for the raiders to cross the sound that the locals couldn't venture out without being robbed. Realizing the Connecticut boats were literally flying the Jolly Roger, in 1781, Governor Trumbull revoked all the whaleboat commissions---not that it changed anything. Raids continued unabated on both sides.

Loyalist whaleboat pulled off similar raids on a regular basis. Loyalist boats raided in fleets, burning mills, villages and twice burning the city of Fairfax. They looted, cut out ships, took prisoners and robbed all that was in reach, continuing up to the last days of the war.

On the Jersey Shore the conflict was somewhat different. Most of the New Jersey whaleboats were built and sailed by individual fishermen or neighbors, banding together to raise the funds necessary to build a whaleboat and pay the privateering bond. Once constructed and crewed, multiple boats often banded together under a chosen leader to defend or to make a specific raid. The population there was mixed, loyalist and revolutionary and communities were always looking over their shoulder; often the populations of villages were pitted against each other. A portion of a deposition made in a New Jersey court reflected on the situation describing an incident in the Toms River area in 1779.

A large whaleboat built by Samuel Brown, of Forked Creek, had been raiding along the coast and up and down Barnegat Bay with some success. In the summer of 1777, his 30-man crew had come up from astern of a medium sized British merchant brig, sailing in light wind. The whaleboat raiders swarmed over her rails before any of the crew could respond and the captured ship and cargo afforded them a hefty profit. Alone or with other raiding crews they had harried British and loyalist interests at every opportunity, capturing enemy gunboats and when possible, defending rebel homes and farms.

Loyalists knew who they were and made the decision to do something about Brown at the first opportunity. In July of 1779, Brown docked his whaleboat in Toms River to make necessary repairs, and with his son, retired to his home a few miles away. A Tory lookout sent word to a waiting party of loyalists. At dusk Brown was alerted by a barking dog; a party of loyalists were fording a creek to approach his farmhouse. Suspecting that he was looking at a group of loyalists out for his hide, Brown and his son ran for it. Both under fire, they headed for Toms River and their boat.

Outrunning pursuers, Brown joined with other captains and soon discovered the loyalist were from Clam Town on the edge of Egg Harbor. Keeping them under observation, the locals mustered a force but before anything could be organized, the Clam Towners set fire to an anchored schooner, and hurrying back to Forked River, they looted and burned Brown's home with all its outbuildings. Mrs. Brown and her children were sent running up the road.

The whaleboaters were not the sort of men to accept such activity passively. Intent on doing something after the attack, local whaleboat commanders met with a Rhode Island privateer Captain by the name of Gray to discuss the Clam Town problem. By 1780, the loyalists residing at Clam Town were growing in number. The town had become notorious, a stronghold for royalists who looted wealth of all sort, including slaves from Rebel Patriots.

Growing up fishing and hunting in the cedar swamps and marshes, the whaleboaters knew every approach and hidden cove. The rumor of a valuable cargo aboard Gray's small ship was spread to Clam Town. Gray, setting himself up as a bait to draw the loyalist out of Clam Town, sailed from Barnegat Bay and out Egg Harbor inlet into the ocean. The plan was for the Toms River whaleboats to lay in wait and when the Clam Towners passed, to follow on their heels, and catch the loyalist vessels between the two forces. The strategy was bold, putting Captain Gray, his men and ship in inordinate hazard. Cunningly, he handled the ship in such manner as to give the appearance that it was damaged. Seeing the Tories overtaking in his wake he made it seem that the ship was pressing on all sail to escape, thus drawing them further offshore. As it was, the pursuing Tory ships were manned and commanded by the same men who had sacked and burned Samuel Brown's home.

As the Tories closed on Gray, Brown's whaleboats were coming up astern of them. Grey suddenly brought his ship about and fired successive broadsides killing and wounding a large percentage of the loyalist crews. Simultaneously, the whaleboats fired and boarded from the opposite side. Those Tories who survived were brought ashore and imprisoned at Burlington. It was a disaster for the Clam Town organization and the remaining loyalists fled the area.

In another instance, the commander of a whaleboat, by the name of Joshua Studson ran out of Barnegat inlet, sailing 50 miles up the coast to Sandy Hook. In the dark of night, he slipped into Amboy Bay and captured a schooner from under the nose of the *HMS Roebuck*. The schooner was sailed a short distance to Middletown Creek and anchored as the elated privateers prepared to celebrate. Unfortunately, *HMS Roebuck* arrived shortly after, retook the ship, burned Studson's whaleboat and for good measure, mills ashore belonging to Captain Burroughs. The moral is, "don't stash the loot where its owners can see it."

Yes, outcomes were far from one sided. The British, guided by loyalist, made forays up rivers into the farming centers of the Garden State, where they seized grain, produce and property as they saw fit. These seized goods were passed on to British ships for transport to the military hub in New York. Rebels used their expertise and knowledge of coastal bays and estuaries to counter the British and recapture property. Also, they often worked to assist Continental forces when asked, by carrying out such tasks as transport, reconnaissance and targeted raids.

Manned 26' Whaleboat

As an example, in 1777, the British had in their possession a large number of American prisoners, some officers from the Continental Army and some civilian Patriot leaders, whom it was British policy to drag from their homes and imprison. Both Congress and General Washington would have liked to foster a prisoner exchange, but unfortunately American forces had neglected to capture any British subjects of enough importance to warrant an exchange. Washington sent an officer to look into possibilities and a plan was hatched to acquire prisoners of the proper influence. The mission was entrusted to Captain

William Marriner, a New Jersey whaleboat privateer. At the army's request, he put together a raiding party of 26 picked men out of a larger group of volunteers for a foray on to Long Island. Accompanying Marriner was John Schenck, a militia officer who having family in the Flatbush area was familiar with it since his youth. Shortly before dusk on June 11th, Marriner's two whaleboats cast off from the shore of Matawan Creek, New Jersey, a point on the southwestern shore of Raritan Bay. It was a gusty day, overcast with winds fresh from the east that were pushing steep waves in from the Atlantic. For the first five miles waves took the whaleboats on the beam, and their crews were continually soaked with spray, until at the mouth of the Raritan River, the boats angled up into the wind, toward and then along, the shore of Staten Island. Night deepened and they held in tight to the beach, so that their boats profile merged with the land, for Marriner wanted them to remain invisible to British patrols. Pitching violently as oars drove the boats up into regularly breaking waves, a few men succumbed to seasickness and were told to suffer in silence or learn to swim.

After a miserable 12 mile pull to windward, the still undetected whaleboats angled across the deep narrows between Staten and Long Island. They nosed up into a patch of thick trees and undergrowth and the men crept ashore. Marriner placed three sentinels to watch over the boats and twenty-five men hurried inland with a list of important Tories who had been marked for abduction. The rendezvous was the graveyard of a small church and from there the raiders divided into 4 groups each with a name and the residence from which a particular individual could be collected. Each team was to hit their assigned home at the same time, grab their man and revisit the churchyard, joining with their companions and returning to the boats in strength.

The four teams of raiders hurried off into the gloom on their separate missions. At the outset, they encountered ill luck. Two of their targets were away at social events but a third, Miles Sherbrook, a rich loyalist broker and particular foe of Captain Marriner was discovered hiding behind a neighbor's chimney. They also discovered Jacob Suydam to be elsewhere but at his home they found billeted an American officer, Capt. Alexander Gradon who was paroled. Taken at the fall of Fort Washington, he was awaiting exchange. The raiding party announced him freed and he gladly joined them. Another team dragged Theophylact Bache from his sleep, into the night and to rendezvous at the church. The abductions had been made in timely fashion, allowing no time for an alarm to be raised. The boats were boarded and cast off immediately. Marriner set a course directly across Lower New York Bay and Raritan Bay for Keyport. The outgoing tide was fair and miserable headwinds were at their back. Sail set, the whaleboats fairly flew and crossed the 12 mile stretch in a rush. A mere 75 minutes to double Keyport and by dawn they were alongside their Matawan dock.

Though a few fish had slipped their net, the leadership was pleased and Marriner was induced to encore performances. Later raids produced so many prisoners that Tories began to move onto Manhattan Island. It could be profitable too, for in one raid Marriner came upon a catch valued at $5,000.00. Patriotic though he was, Marriner didn't limit himself to government missions. When he was not raiding inland, he was launching his whaleboat crews against merchant shipping and was a famous pain in the British posterior for a few years. On dozens of occasions he took small sloops and barges transporting goods along the Jersey coast, Raritan bay and lower New York Bay, even up into the Kills.

One incident published in outrage by New York papers; a number of small British ships had taken refuge from a storm in the shelter of Sandy Hook.

Marriner had quickly organized a raid; with several whaleboats. In the dead of night, he led the whaleboats down the bay, pouncing on the anchored vessels as their crews slept. The sloops ended on the beach, stripped and burned. Marriner raised sail and made off with the schooner. On

inspection of its hold, they found an expensive cargo that brought each of the raiders a small fortune in prize money, all for a single night at sea.

Marriner's most audacious raid began on the night of April 18th, 1780. The British frigate *Galatea* had overhauled and captured a lovely little privateer brig, the Blacksnake of Rhode Island. Crewed by a prize crew of twenty of the frigate's sailors, she now lay nestled under the protective wing of a three decked British ship-of-the-line, the powerful *HMS Volcano*. Marriner found her beautiful; she had caught his eye and now he lusted for the haughty little brig. Well, a man in the throes of lust is often known to become witless.

Finding only nine Brunswick men available on short notice, barely enough to man a single small whaleboat, he sailed six miles down the Raritan river to the bay. The westerly breeze was light astern as they crossed Raritan Bay. They rowed the twelve miles in two hours. Off the tip of Sandy

Hook, they muffled the oar locks as they approached the anchorage. Under cover of a light mist, they closed on the brig. There was absolute silence. The raiders were outnumbered more than two to one, but they were awake, and the prize crew was not. Swarming aboard, a lone lookout was subdued, and the crew locked below decks. No resistance and not a sound to give them away. Immediately, the anchor cable was cut, and a bit of sail set, only enough for the Brig to begin ghosting out of the anchorage on a beam reach. The *Blacksnake* began to slip away, noticed by neither the three-decker nor the shore guns.

Marriner and his nine men were out of sight of land when dawn broke and what did first light reveal but the armed schooner *Morning Star*, northbound with a 33-man crew. Brazen to the core, the whaleboat men sailed down on the schooner guns blazing. The *Blacksnake's* bold attack was daunting, and the *Morning Star* surrendered. When the brig laid alongside, the schooner captain realized he outnumbered the whaleboat-men three to one. He urged his crew to attack but the whaleboaters responded with such ferocity that in seconds several lay dead with more wounded and he surrendered a second time.

The nine whaleboaters, who 12 hours before had been in possession of a 28 ft boat were now in possession of two ships and fifty unhappy prisoners. They were tired and making port with their prizes was critical. By 10 AM, a backing onshore breeze allowed for a direct course to Cranberry Inlet at Toms River. They had come to safe anchor before dark, and on a later date the prize ships were sold at auction.

Back at collecting hostages for the Continental Congress a short time later, he was ambushed and captured on Long Island. Paroled by the British, he returned to New Jersey and opened a tavern, but Marriner had colleagues, as enterprising and even more intrepid than himself. One such was another Brunswick man, Adam Hyler.

Hyler, who was German born, had been pressed into the British navy as a youth and although he had learned seamanship, he'd not been pleased with the experience. He was a citizen of Brunswick, with a modest fleet of trading vessels carrying cargo between Brunswick and New York, when revolution broke out. He was originally associated with Marriner's operations but with Marriner's compulsory retirement his exploits came to the forefront, easily surpassing those of Marriner.

He was not reported to be a tall man but rather stocky and built like a guerilla. His name began to appear in the newspapers shortly after cessation of Marriner's activities and he speedily became associated with astonishingly and incessant action. For a solid year, Adam Hyler and his whaleboat men drove the enemy to distraction.

He cut out small ships from anchorages, led whaleboat raids up rivers and inland capturing entire loyalist militia units. Like Marriner, he began carrying out raids on Long Island to acquire hostages for prisoner swaps. When on October 5th, 1781, intelligence of a convoy of merchant vessels at Sandy Hook reached him, Hyler quickly contacted whaleboat crews in the area, six joined him. He boarded his privateer sloop, the *Revenge* and coasted downriver, exiting at South Amboy and into Raritan Bay at dark. The night was overcast as the *Revenge* sailed the 12 miles east toward Sandy Hook.

Like Marriner had sixteen months before, Hyler must contend with the massive guardship and shore guns. He, unlike Marriner wished to capture not one ship but five vessels: three merchantmen, and two small warships. Odds of success were the most unfavorable that whaleboat men had ever been challenged by.

Nearing the anchored flotilla, *Revenge* was brought up into the light wind and with jibs aback, gently drifting leeward her shape obscure on the black water. Silently, two whaleboats with muffled oars set off to reconnoiter the anchorage. Hyler and the others waited in what must be described as a tense atmosphere. The intelligence was heartening. The crews of the merchants were ashore, leaving only a watchman. The two warships were manned but so unattentively the whaleboats had passed under their sterns unnoticed. They had not probed so near the massive guardship, but aboard her, not a soul seemed to be alert but rather benumbed with sleep.

Hyler wasted no time deciding to pounce on the anchored ships with all dispatch. He directed one whaleboat crew to board each of the three deserted merchantmen. The other whaleboat crews would board the smaller warship, while he and the men of the *Revenge* would lay alongside the stronger one.

Soon, all were in position and helm down, *Revenge* glided toward the anchorage; she groped through the dark, discerning which of the indistinct shadows was her chosen victim. Shadows grew, took shape and *Revenge* adjusted direction, closed the gap until at 45 degrees, her bow gently touched the stern rail of the British vessel. Ten men leaped aboard before the first grapnel caught. The sleepy deck watch had hardly the chance to squawk out an alarm but the rush of pounding feet managed that for him.

The newly woke crew, grasping whatever weapons came to hand, attempted to gain the deck. The few that actually succeeded in reaching the deck, found themselves overwhelmed and dove over the side to extend their lives. The remainder found the hatch guarded under threat. Any head that rose above the coaming was guaranteed to be lopped off. Not enticed to tempt such a fate, the crew remained below, and the ship was carried, in less than a minute, without the loss of life.

Aboard the other ships, the whaleboat men had succeeded with like efficiency to triumph. The smaller privateer had been carried, only a few of her crew escaping in the ship's yawl. The merchantmen had been virtually deserted. Aboard them, the whaleboat men found only the sleeping watches and, in the cabin of one, a wife with four terrified children hanging on her. A few had escaped ashore, though and would give alarm to the shore batteries. Time was running out and the raiders would have to move fast.

The highest valued cargo from the captured vessels was swung across the deck and dumped into the *Revenge's* hold and open whaleboats. As they worked, shore guns began to fire as the fort's swivel gun crews attempted to hit the ships. They found the ships beyond their range but the noise they made had woken the officers of the guard ship who were certainly trying sort out what was going on. Already lights could be seen in the big ship's open gun ports. Adam Hyler knew when it was time to depart.

Sparing the merchant ship with the woman and her children, they set fire to the other two. They let fly all sail, and with whaleboats in tow, they faded into the enveloping darkness. Astern, burning vessels lit the sky, and the big guns aboard the guardship finally thundered. Sandy Hook Bay was pocked with frothy spouts of water as the area was peppered with plunging round shot. The little flotilla had moved off fast though and was all but invisible in the night.

The whaleboat raiders entrance into New Brunswick the following day was celebratory. The New Jersey Gazette reported delightedly that the whaleboaters arrived with prizes: prisoners, sails and cordage, plus valuable cargo hoisted from the captured vessels. Hyler continued his raids but died after an unfortunate accident ashore in September of the following year.

There were of course many hundreds of whaleboat raiders. Others like Captain Dickie, or like Hallack, and Solomons who was involved in the Sags Harbor raid. The raid, led by Col. Meigs crossed twenty miles of open seas in whaleboats, made a bayonet attack on a British garrison. Ultimately, the raid would cover over ninety- four miles in twenty-four hours, most taking place at night. They captured ninety some loyalist soldiers, killed eight, didn't lose a man killed or wounded. The raiders torched tons of British stores and assaulted a man-of-war of 12 guns. This ship was set ablaze with a dozen other British ships moored in the harbor. The raid was considered not only a success but a miracle. They were only following in the wakes of thousands of years of other sea borne raiders.

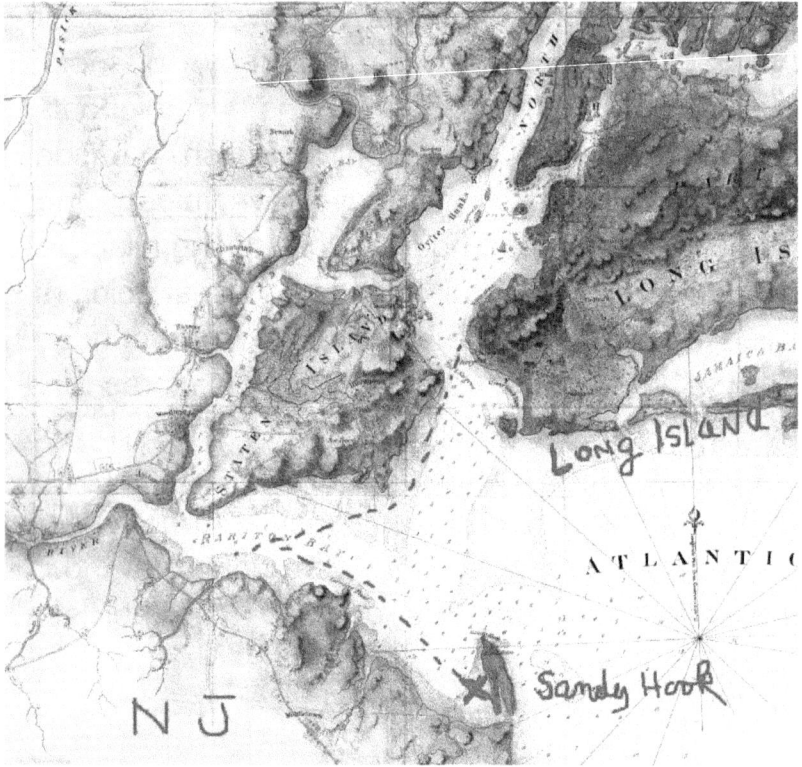

The Privateers

A merchant ship with her cargo holds filled, her passengers idling on deck, sails along, bound for a distant port. Suddenly a lookout calls out from the masthead, *"sail ho"*. Suspicious, the merchant vessel sheers off, cautiously seeking to avoid the stranger but the unknown vessel turns with them and narrows the distance with eerie speed. Soon, decks packed with armed and ununiformed men, the vessel is surging alongside. "Pirates," the quaking passengers scream.

Wait a moment now! Maybe not that bad folks.

Suddenly, the ship runs up the flag of a nation presently at war with theirs. A canon fired; rounds shot skipping over the water ahead of the merchant throws froth into the air. Not pirates thank heaven. Yes, bad but not that bad; **it's a privateer and they have rules**.

This US Revenue Cutter was a type favored as a privateer.

What's the difference? Well the easiest word to explain it to modern ears is bounty hunter. A pirate or when on shore an outlaw is out to rob everybody. A bounty hunter is holding a piece of paper from the government, with the name of a culprit, a picture, and the promise of a reward for apprehending him. He's not supposed to go arrest or shoot anybody else. The piece of paper licensing the privateer to attack his nation's enemies was called a *Letter of Marque* or Letters of Marque and Reprisal.

Chesapeake built Fore Topsail Schooner

Privateers had loose perimeters, the warranted actions outlined or described on the letter of Marque were legal. The term "privateer" describes the ship, its captain and any member of the crew, all of which are civilians but, in this capacity, they represent a nation or state. The ship's owner or owners were required to post a bond before being issued the Letter of Marque which commissioned them for a certain period of time, or for the entire period of hostilities. Privateering allowed sovereigns to raise revenue for war by mobilizing privately owned armed ships and sailors to supplement state power. Specific responsibilities were described, with instructions to capture or destroy vessels of the enemy nation on land and sea, as well as take possession of their property.

The Privateer didn't get to keep it all. Captured ships and cargos had to be taken to port, condemned, and sold. The state and all involved, then got their prescribed share and every country commissioning privateers, operated a prize court which judged every claim to "a seized vessel and its cargo". A court would refuse a claim and release a ship if it determined that it belonged to a neutral country, or if it's cargo was owned and consigned to neutrals. A major benefit of the policy for the state was that money was paid into the states treasury by a civilian navy that cost them nothing. The nation and privateers alike, benefited from this arrangement. Unfortunately, opposing nations often refused to treat captured privateers as prisoners of war but then prisoners of war didn't have it all that good anyway, particularly in ancient times.

Most scholars would tell you that Henry III of England granted the first Letter of Marque or Reprisal in 1243. Actually, we need to go back another 1500 hundred years to the Illyrians. Queen Teuta of the Ardiaei, who inherited the kingdom of her husband, Agron in 231 BC. She then acted as regent for her young stepson. The territory of Illyria stretched from the northern borders of Greece and Macedonia to the Northern Adriatic Sea, encompassing much of modern Albania, Montenegro, Bosnia and Croatia. The Illyrians were traditionally pirates and piracy was an important part of the economy. Entering into hostilities with most of her neighboring states in 231, Teuta granted every ship in her kingdom a license that gave them permission to attack ships from other hostile city or states including some who didn't know they were hostile. Effectively they weren't pirates but privateers, endorsed by the sovereign Illyrian state.

Rome's interest was directed toward conquests on land and as a policy, Rome ignored problems at sea. This changed when Illyrian privateers seized a fleet of Egyptian grain ships sent to supply their legions. Two envoys were sent to Illyria to complain. One did so with such disrespect (Roman men were what has been called *MALE CHAUVENIST PIGS*) that Queen Teuta had him fed to the fish. The outraged Roman Senate then declared war but that's another story.

The Ottoman Empire Commissioned Privateers, (Corsairs) in a similar manner. Same thing, private ship, state sanctioned attacks with a percentage coming back to the sovereign. Forms of it took place in the Baltic and North Sea and is mentioned in texts written during the Hanseatic League's 400-year tenure as a trading power confederation. The real beginnings of Privateering as we define it, began during the Elizabethan Era.

Though no official state of war existed between England and Spain, a religious war did. Queen Elizabeth I issued letters of Marque to the captains, who came to be known as her Sea Dogs, commissioning them to attack and loot Spanish ships and lands.

The Elizabethan Sea Dogs.

Small English Ship of middle 16th century

Although there were many others, the most famous were Hawkins, Frobisher, Raleigh, Newport and Drake. Phillip II of Spain did the same but without the same success.

Sir John Hawkins began a career at sea in his youth. He recognized early that great profit was to be made, if he developed an African Slave Trade. The trade route he worked out came to be known as the triangle trade, but his slaving activities brought him into direct conflict with Spanish trade restrictions in their American colonies. English trading rights with Spain and Spain's possessions were established during the reign of Henry VII. Spain now stated that her new world colonies were not included in these treaties. Thus, during what the Spanish now considered an illegal trading voyage, Hawkins' fleet was trapped and all but two ships were destroyed in Vera Cruz. The two small ships that escaped were Drake's *Judith* and Hawkins' *Minion*. Hawkins eventually arrived back in England with only eleven men alive on arrival partly due to starvation. Drake had reached England a month earlier in slightly better condition. The Spanish inquisition sent many of those captured there to the galleys and to be burned at the stake as heretics. Hawkins spent a long life making the Spanish pay.

Sir Francis Drake, known as 'El Draque' ('the Dragon') by the Spanish, attacked the Spanish settlement of Nombre de Dios Panama in 1573 and captured a silver caravan. Drake later completed the circumnavigation of the globe between 1577 and 1580 CE during which he raided Spain's South American settlements. He captured the treasure Galleon *Nuestra Senora de la Concepćion*, and a week was required to transfer the gold to Drake's *Golden Hind*. He plundered other ships, on the west coast and returned to England at the end of a two-year nine-month voyage. The loot amounted to £600,000, more than double the entire annual revenue of England and the pleased Queen knighted him as a reward.

Drake continued privateering for another decade, making daring raids in the Cape Verde Islands, Hispaniola, Cartagena, Florida, Cuba and San Juan. In 1587 CE Drake proved the practicality of privateers in national defense, with an assault on Cadiz Spain where he captured six ships and burned 31 others, destroying materials destined for the construction of Philip's great Armada. **Martin Frobisher** was, like others, as much explorer as privateer. He did take Spanish ships but perhaps, seized more from the French and he led fleets on several voyages searching for a route thru the North West Passage.

Sir Walter Raleigh, the privateer wanted to have a base for cruising against Spain's Caribbean shipping and organized three expeditions to form a colony on the coast of North America in the 1580s. The colonist of Roanoke Island, Virginia disappeared but the one at Jamestown took. Later, Raleigh went on two expeditions searching for the fabled city of gold, in what is now Guyana, in South America in 1595 and 1617.

the English privateers. On a 1596 raid in Spain the old Sea Dog, had his place in a fleet that destroyed 50 ships and for the second time, sacked the port of Cadiz. The crown's share alone amounted to £80,000 worth of goods. Still he angered the king and ended his days in the tower writing his history of the world.

The 3rd Earl of Cumberland, Sir George Clifford made a fortune privateering against the Spanish in the Caribbean, and in 1592, even captured San Juan, Puerto Rico's, Fortress, El Moro, the citadel protecting San Juan and its harbor. Alas, in gambling he was less successful and lost it all betting on the horses.

17th century privateers

Of naval forces at sea in the 1600s privateers comprised a major part. In the Anglo Dutch trade wars, English privateers preyed on the trade of which the United Dutch Province's economy was hinged on. Over 1,000 Dutch merchant ships were taken in the first war. During the following conflict, it was English shipping that suffered as Flemish and French privateers in the service of the

Spanish Crown, seized 1,500 English merchant ships. British trade, whether coastal, Atlantic, or Mediterranean, was also plundered by Dutch privateers in the following conflicts known as the Channel Wars. Sides changed with such regularity that privateers could hardly keep the national flags straight.

Piet Pieters Zoon Hein, a Dutch captain, was a dazzlingly success as a privateer and captured an entire Spanish treasure fleet. He and other Dutchmen were against the Spanish but so much wealth flowed from the Americas to Spain that they could never seize it all. They never captured more than a small fraction.

Providence Island. Most old movie aficionados and pirate buffs have heard of Jamaica, Hispaniola, Tortuga, Nassau, Portobello, etc. but few are aware of a colony established by Puritans, supposedly for the purpose of agricultural that really was planned as a base for privateering. Providence island, stumbled on by the English in 1628, was unique.

Middle 1700th Century, Armed Dutch Merchant Ship

The Channel Trade wars of the 1600's

Providence, The island Colony of Privateers

The 2nd Earl of Warwick had sent three privateering ships to the Caribbean in 1627, during which time they discovered San Andres. They decided to raid from there while residing ashore and planting crops. One ship discovered Providence 45 miles to the north, and returning to England, by way of Bermuda, informed Governor Bell. Hearing the news, Bell wrote a relative in England of the profit to be had with a colony and privateering station, located in the midst of the Spaniards. Philip Bell sailed to Providence in 1629, with a party to see it for himself. A company was formed, and a settlement started in November of 1630.

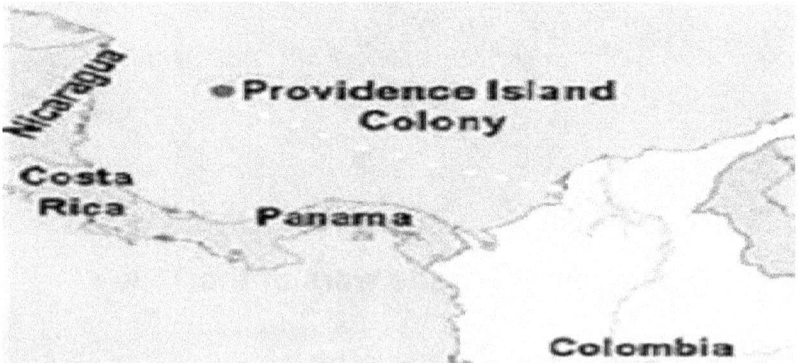

By 1635 the population was recorded as 600 white men, 70 women, 100 blacks but the Spanish had no hint of the Providence Island colony's existence. That changed when they caught some Englishmen lurking about the isthmus city of Portobello. A force was sent to destroy the colony, but the Spaniards were repulsed and sent packing. On news of the attack on his colony, Charles I of England issued letters of Marque to the *Providence Island Company* authorizing raids and seizures in retaliation for the Providence incident and for an earlier attack on his colony of Tortuga.

Early in 1636 the Company sent a ship, the *Blessing,* with Captain Robert Hunt who was to be the new Governor of what was now a Company base for privateering. Hunt's primary interest was the Spanish treasure fleet sailing from Panama to Havana in 1637. Hunt accomplished his task of organizing privateering and was replaced in 1638. Nathanial Butler, formerly Governor of Bermuda, was an expert in fortifications and his task was to fortify Providence. This was completed in 1640 and he departed, deputizing the governorship to Captain Andrew Carter.

Meanwhile, pressure continued to build between Spain and England, which were still technically at peace. In 1640 the Spanish Ambassador in London stated that ships bearing gold, silver plate, emeralds, pearls, jewels of every sort, and many other valuable merchandises belonging to subjects of his Catholic Majesty, the King of Spain had been seized by ships of the Earl of Warwick and other ships from the island of Providence. In a time of peace, his countrymen were being robbed, impoverished and murdered, all in league and friendship with Charles I.

The situation having become intolerable, the Governor and Captain-General of Cartagena sent a force of 1000 made up of six hundred armed Portuguese, 200 black and mulatto militiamen, and 200 marines from the ships. The troops were landed on the island and engaged it a ferocious assault which was turned back. Defeated, some of the Portuguese and Spanish officers surrendered with the assurance that their lives would be spared. Captain Andrew Carter, acting Governor of the colony, ordered all thirteen Portuguese and Spanish prisoners put to death.

Outraged, King Phillip III of Spain instructed that they were to be avenged. The Spanish acted definitively, sending another expedition from Cartagena to Providence with seven large ships, accompanied by pinnacles, 1400 soldiers and 600 sailors. General Pimienta, personally led the assault on the morning of 24 May. The Spanish and Portuguese troops quickly gained control, and the English negotiated a surrender. The Spanish and Portuguese recovered gold, indigo, six hundred black slaves, and stores of booty valued at 500,000 ducats. Over the following century the island passed from nation to nation but never again became a real threat as a privateer base again.

Captain Henry Morgan, later Sir Henry, was indeed a privateer but far more brutal than the day to day pirate of his time and sailed full sized ships of the line. Based in Jamaica and under the protection of the governor Sir Thomas Modyford, Morgan needed cover for he was disposed to cruelty, and torture when information concerning gold was needed. He operated against ships and fortified cities with equal zeal, used priests and nuns as shields. He assaulted and sacked the city of Panama, gained enormous amounts of booty and promoted rape, torture and every form of cruel excess desired by his men and he was loved. Eventually, he became Governor of Jamaica and a pillar of island society; today, there's even a popular rum distilled that bears his name.

Bermuda Privateers

The isles of Bermuda were first reported in 1505 by a Spanish explorer, Juan de Bermudez and were included on Spanish charts. Spanish and Portuguese ships visited the archipelago to water and replenish stores, but frequent storms, dangerous reefs, and rumors of demons and spirits gave the place a bad reputation. The archipelago became known as the 'Isle of Diablo' and neither country was tempted to settle, leaving it uninhabited for another century.

England had by this time become interested in colonizing North America but the first two attempts at establishing colonies had failed. King James I, determined to correct this, granted a Royal Charter to the Virginia Company causing the Jamestown, Virginia colony to be founded in 1607. Two years later, a fleet of seven ships, bearing food supplies and hundreds of new settlers departed England to relieve and expand the struggling colony. The fleet was scattered by a storm. George Somers' flagship, her hull strained in the tempest was leaking faster than she could be pumped. Sighting Bermuda, he determined to drive her ashore. The *Sea Venture* was beached on a reef and her crew and 150 passengers landed in the ship's boats suffering no loss of life. Somers dispatched a longboat to bring rescuers from Jamestown, but it vanished. Somers then began the settlement of Bermuda and once it was organized, he ordered two small ships built. They were to be the *Deliverance* and the *Patience*. After 10 months the ships were complete and in May of 1610, the survivors of the *Sea Venture* continued on to Jamestown, leaving two men to maintain English claim to the Bermuda Isles.

The population of Jamestown was found to be starving and Somers returned to Bermuda to secure provisions but died en-route. The captain of the *Patience* made for England instead and in 1612, the Crown began the intentional settlement of Bermuda. Additional settlers came from other English colonies in the Americas, indentured servants, some from former Spanish colonies in the West Indies, now in English possession. There were always the victims of shipwrecks, enslaved Irish prisoners, victims of the attempts to eradicate the Irish, and also American Indian slaves. African slaves arrived only when a slave market was established for bond slaves. Obviously, Bermudians colonist were racially diverse. With only 21 square miles of land and no natural resources other than the Bermuda Cedar, the colonists found agriculture, particularly tobacco to be economically impracticable. To feed themselves they began to plant food crops, raise cattle, pigs, chickens, turkeys and to grow guavas, oranges, lemons and other fruits. To make cash, they invested themselves in the maritime trades and were soon able to provision ships that put in, with need of stores. Supplying ships was good business but It was ship building though that lifted the economy and made the Islands famous.

A Dutch shipwright, who had been castaway on Bermuda in 1618, was hired by Governor Butler to build needed boats a year later. In the first three years, there were five times as many vessels on the island and the Dutchman had trained numerous apprentices. More importantly, he developed a radical new design, based on a type of vessel used in the Netherlands, **the Jacht**. Modified, the Jacht became the Bermuda sloop, a vessel destined to alter the direction of ship design in the days of sail.

Bermuda sloops, through built light, were strong. With deep fine hulls and sails rigged fore and aft, rather than square sails, they could point up-wind at angles other ships couldn't come near to matching. Faster to windward than anything afloat in an era of square-rigged ships, the sloop's ability to point high into the wind was an enormous advantage to those wanting to catch or evade another vessel.

At the beginning of the seventeenth century at least two dozen ships a year were being built, plus additional small craft. Privateers, who depended on speed and maneuverability, treasured them. They were used not only by English but were mentioned in Caribbean conflicts between the Dutch and French around 1700. They became even more popular during the War of the Spanish succession and continuous little conflicts like King William's War and King George's war and others. In 1729 the war of Jenkins Ear began and flowed into the War of the Austrian Succession lasting until 1748. Within 8 years it began again with the

Seven Years' War, involving European colonies all over the globe.

It was a great time to sell fast ships and sailing them was even a greater money maker. At first the Bermudians only dabbled in the privateering part. A few Bermudian sloops were commissioned as privateers during King William's War, but with unimaginative captains, most hardly broke even. On the other hand, a 14-gun Bermuda-built sloop, purchased and commanded by the Spaniard, Captain Don Francisco Lorenzo was highly successful. He became the most famous Privateer of the decade.

Three mast Bermuda Sloop

The Bermudians began having more success at privateering at the start of hostilities in 1739, the War of Jenkins' Ear. The Bermuda-built, *Popple* captured a Spanish ship and cargo valued at £7000 plus 8,000 pieces of eight. Within a year, Bermuda had licensed double the number of privateers of any mainland colony. By the end of the war, Bermudian vessels were involved in privateering everywhere. Even sovereign states and colonies acquired Bermuda sloops for coastal defense. The Royal Navy had a number of traditional Bermuda sloops built, even a few large ones were built as frigates. The design was influential in American shipbuilding and though during the revolution many Continental privateers were built in Bermuda, their type was modified in the famous Baltimore clippers built in the Chesapeake Bay.

The islander themselves, continued to sail as privateers until the end of the war of 1812, during which, Bermudian privateers captured 298 American ships, amounting to 20% of the nearly 1,600 vessels captured by combined British naval and privateering ships. The captain of a US naval vessel who was given orders to depart Boston Harbor and deal with a pair of Bermudian privateering vessels. They had been seizing vessels which had so far escaped the British navy. He returned without success, and angrily reported that the Bermudians had sailed their ships two miles for every one of his sailed.

Privateering Laws Change

By 1700 laws regarding piracy and privateering had changed somewhat, altering the definition of piracy from treason, to a crime against property. The British Piracy Act of 1717 defined a British privateer's allegiance as being to Britain and thus overrode any contract such British subject made with a foreign sovereign that might be providing him with a commission. Other nations soon adopted the policy, and this placed Privateers under the jurisdiction of their own nation.

But what a nation was and who represented it was not always easy to determine outside of Europe.

Privateering was also a factor on the Far East.

Kanhoji Angria, for instance, who was given the title of admiral in 1698 became an important naval figure in the Indian Ocean. Under that authority, he was responsible for the Western coast of India from Mumbai to Vengurla. Until his death in 1729 he maintained absolute control over the coast. In the face of European incursions, Angria was the naval arm of the massive Maratha Empire and he not only out did the Dutch, British, French and Portuguese but dealt with local threats and rivals ashore.

His business was seeing trading fees were paid for ships entering his waters and the Europeans didn't choose to cooperate. He began his career by capturing ships and issuing what amounted to Letters of Marque. At its zenith, in the Maratha fleet put together by Angre, there were never more than 80 ships, many of them under 70 foot in length.

With an unmatched mind for naval tactics and strategy, he outwitted his adversaries at every turn, though. He captured merchant ships even inside harbors. Sank the frigate Bombay in a sea battle and within two years had the deference of the colonial powers. The British realized he could take on all but the largest European warships. Angria was never concerned with influencing trade and traffic in other than the coastal waters he was commissioned to protect but, in those waters, he established a formidable authority in the interest of the Maratha Emperors.

Muslim Sulu Sultans of the Sulu Archipelago, a group of islands (now known as the Philippine islands, extending south to the Sulawesi Sea) in the regions bordering the South China Sea, held only a shaky authority over the local Iranun slave-raiding communities. These peoples were the product of religious colonization by Iran centuries past. Only a Machiavellian tangle of military and political associations controlled unauthorized raiding. The Sultans understood that raids and piracy conducted against European ships and colonial settlements would provoke a war they had no hope of winning. As a result, they took advantage of the competing European nations. Always at war with each other, it was easy to make short term alliances with Europeans and issue Letters of Marque, giving legitimacy to the bloodthirsty Iranun raiders and at the same time exerting some control over them. A merchant seaman's life could be quite chancy in far east.

American Revolution and wars of the Napoleonic Period.

In 1776, there was little difficulty in obtaining a commission as a privateer for a British subject or a Continental. With the passage of an act on March 23, 1776, the Continental Congress formalized the commissioning process, and uniform rules of conduct were established.

The act began, *"That the inhabitants of these Colonies be permitted to fit out armed vessels 1st to cruise on the enemies of these United States."*

From almost every port, vessels of every size and type were armed as privateers; ships the size of frigates down to whale boats. Two-masted schooners brigs and brigantines were favored, particularly Bermuda sloops and the fast, new Chesapeake built schooners. Owners of privateers had to post monetary bonds to guarantee their proper conduct under the regulations and captured vessels and cargos had to be delivered to special courts which would examine the legality of the seizure. Congress would issue 1,697 Letters of Marque during the Revolutionary War. Individual states issued an additional 1100, plus those letters issued privateers commissioned by Benjamin Franklin in Paris. Some privateers had beginners' luck. One American privateer, the *Rattlanalee* cleared $1 million from a single short cruise in Baltic waters. She was captured on her next cruise. Oh well!

Out of Beverly Massachusetts, one of the more lucky and active vessels in the first months was the 18-gun, ship-rigged *Pilgrim*, owned by the Cabot brothers and captained by Hugh Hill, a man known for his dislike of the British government. During the siege of Boston, Hill captured and delivered the British supply ship *Industry* to Washington's forces. On cruises to waters around the British Isles in the first three years of the war, he captured dozens of merchant ships, earning a reputation a nickname, "*the scourge of the coast*". Described as a man of unusual height and strength, Hill was also known for a commanding demeanor, cool nerve and crafty ways, traits that made him a legend in privateering circles.

On an occasion when the *Pilgrim* was overhauled by a British warship, she was flying the British Flag as a ruse. Boarding, the British man-of-war's master announced that he was in search of that "*notorious privateer Hugh Hill.*"

"*As am I,*" Hill said, while presenting the forged papers of a British privateer. After a brief

and affable conversation, the British captain disembarked and sailed away. Later with a full crew aboard *Pilgrim*, the two ships met again. This time Hill raised the American colors and commenced firing. After a brief engagement, the British vessel surrendered. A smiling Hill then introduced himself to the stunned British captain.

After capturing the British ship *Mars*, in 1781, Hill returned to Massachusetts. The *Pilgrim* passed to the command of Captain John Robinson and Hill took command of the purpose-built privateer, *Cicero*. A ship-rigged vessel of 200 tons, *Cicero* was armed with 10 long nine- and 6 four-pound guns and 2 twelve-pound carronades, all manned by a complement of over a hundred. Placed under Hugh Hill's command, it was obvious *Cicero* was intended to carry the war to British shipping in an aggressive manor. On cruises to the Caribbean and west coast of Europe, the *Cicero* captured a large number of vessels, making her owners another fortune by wars end.

In Salem, the ship owner most often associated with successful privateering during the Revolutionary War was Elias Hasket Derby. During the length of the conflict Derby equipped 85 ships. These privateers made over 110 voyages and seized 144 enemy vessels. He put the earnings into trading ventures to the far east after the war and became the new nation's first millionaire. A truly gutsy captain who sailed out of Salem, was Jonathan Haraden of Gloucester. Sailing on a series of Salem privateers, Haraden captured 60 enemy vessels and captured more than 1,000 men. One fight was legendary. In June of 1780 off the port of Balboa, Spain, while in command of the privateer, *General Pickering*, he tricked a big schooner into surrendering. Before the prize crew could get schooner *Golden Eagle* into port, it was recaptured by the *HMS Achilles* on the following morning.

Haraden was not about to lose a fat prize. Even though the *Achilles* had triple the manpower and guns of the *General Pickering*, Haraden took his ship straight at it. Watched by Balboa's Spanish population he closed, coming tight under *Achilles's* bows. The smaller American ship peppered the hull and rigging of the British ship with canister and bar shot while the *Achilles's* cannonballs flew harmlessly above the hull of the *General Pickering* which sat much lower in the water. Eventually, unable to board or seriously damage the *General Pickering,* the *Achilles* retired. Haraden's crew then recaptured the *Golden Eagle.* When he landed at Bilbao, Haraden, for his courage, was paraded through the city by thousands of admiring Spaniards.

Not every privateer came out on top.

A merchant by the name of Clement of Saint Dominque, had the Brig *Guest* fitted out as a privateer. She was armed with twelve guns, sixteen swivels, and manned by one hundred men, all a mix of French and Spanish with the exception of her captain, Commander Edward McKaller of New York. She sailed from Môle St Nicolas in October of 1777 and was patrolling near Montego Bay on the north coast of Jamaica by October 14 when she sighted a square-rigged ship. McKaller ordered *Guest* brought about and the vessel was chased, captured, and had a prize crew placed aboard. She was sent off toward Môle Saint-Nicolas and the *Guest* resumed her patrol capturing a Jamaican shallop later in the day. By chance, the fore topsail schooner, *HMS Racehorse,* a vessel of ten 3-pounders, eight swivels and thirty-eight men, was in port making a repair. Her master received a report of a privateer cruising off the harbor entrance. Acting Lieutenant Charles Jordan sailed at once, sighted a strange sail, and chased it until darkness fell. During the night Jordan steered to the northwest and dawn found the shallop close aboard and discovering it was a prize, disabled it and laid course for the *Guest*. Both ships beat to quarters.

The vessels were in pistol range before 8 AM and closed to fifty yards, both maintaining a steady canon fire. After more than an hour a tangle of rigging caused them to lay hull to hull and Jordan let the crew of the *Racehorse* aboard the American taking the ship after a short fight. Casualties aboard Racehorse were, one killed and eight wounded. *Guest* suffered eighteen killed and forty wounded, of those many more were expected to die, and the rigging of both ships were seriously shot up. Not a good outcome for the privateer.

Often Captains had trouble with paperwork, bureaucrats or politics. Gustavus Conyngham had the worse of them all.

At the beginning of the American Revolution, Irish born patriot, Gustavus Conyngham was sent to France in 1776 to obtain certain military supplies for the troops. Arriving in France, he became friends with Benjamin Franklin and was able to obtain a naval commission from Congress. He was provided with the lugger *Peacock,* an armed vessel he thought had been purchased and fitted by order of these commissioners. Luggers, were small but seaworthy vessels and relatively fast. In the English Channel they were used mostly for smuggling and privateering. Conyngham renamed *Peacock*, the *Surprise* and recruited a crew of American sailors. Men who had been detained for various reasons along the coast of France and Belgium.

Conyngham immediately proved to be an able and very skilled captain, seizing on average a ship a day, he was soon well known to both the British and the Americans. The French maintained a position of strict neutrality and did not allow for a privately owned ship, a privateer, to make forays from their ports. Conyngham considered *Surprise* to be a Continental Navy warship though and his crew were to be governed by the regulations made for Seamen in the Continental Navy. Conyngham, his officers and crew, all believed they served on a "Continental Vessel". An American Commissioner stationed in France, William Carmichael, warned Conyngham in July of 1777 to curtail his commerce raiding and do nothing to upset the French Ministry for a while. His operations had become a sore spot and more of the same would get him in trouble.

Continental Naval Ship, the Lugger Surprise, 1776.

Instead, he was to prepare Surprise for a voyage to America with dispatches. Not taking the hint, Conyngham went out for another raid on British shipping, capturing the brig *Joseph* and the mail packet, *Prince of Orange*. When the prize vessels, escorted by *Surprise*, approached Dunkirk's harbor, they were interdicted by two patrolling British Navy ketches. The Ketches rammed *Surprise* multiple times in hopes of provoking Conyngham into firing on them but now in French territorial waters, Conyngham maintained his respect for French neutrality. Instead. he sued the British for damages in the French court.

Bad idea!

The Surprise is confronted by a British naval ketch.

The 1713, the Treaty of Utrecht between France and England explicitly closed the ports of either power to the enemies of each other. Stories about Conyngham's seizure of British vessels had appeared in English newspapers and were an embarrassment to the crown. Thus outraged, the British Ambassador vigorously protested in Paris objecting to the seizure of his Majesty's vessels from a French channel port. British vessels had appeared in English newspapers and were an embarrassment to the crown. Thus outraged, the British Ambassador vigorously protested in Paris objecting to the seizure of his Majesty's vessels from a French channel port. To emphasize the point, the British sent the 18-gun sloop *Ceres* to blockade Conyngham and his entourage, which left the French no choice. They arrested Conyngham and his crew and in doing so, seized *Surprise* and confiscated his Continental Navy commission and documents.

The British sent two additional sloops of war to Dunkirk which, along with *Ceres,* were to extradite the Irish expatriate and his crew to England for trial. Benjamin Franklin had fostered sympathy for the American cause in Louis XVI court though and was able to gain the release of Conyngham, and his ship. A new Continental Navy commission was delivered to Conyngham dated May 2, 1777, dutifully signed by John Hancock and validated Conyngham's previous service in the Continental Navy. Benjamin Franklin had his agent sell the *Surprise* and purchase a speed-built cutter named *Greyhound* which according to its papers was resold to an Englishman by the name of Richard Allen. All looked correct and aboveboard until the *Greyhound* departed Dunkirk.

Naval Cutter of 14 Guns

Astonishingly with a splash of saltwater Richard Allen morphed into Gustavus Conyngham who renamed the vessel **Revenge.** Assuming his appointment as a Continental Navy officer made the cutter *Revenge* a Continental Navy ship, Conyngham went on a cruise against British Commerce. *Revenge* armed with 14 cannon and 22 swivel guns and a crew of 110 veterans created havoc with British shipping, destroying and seizing many smaller vessels.

Reports of shipping losses caused prodigious anxiety in London. The maritime insurance rates increased by twenty-eight percent, and not only did the British traders asked for the protection of war ships, they to be safe, they switched cargoes to French and other neutral nation vessels.

Perhaps, lacking in political acumen, Conyngham landed his prizes at Dunkirk once again, infuriating the British ambassador for a second time.   The French ministry ordered Conyngham's prizes returned to the British owners and imprisoned the American and his crew. Even the agent who had procured the ship was incarcerated in the Bastille, thus the agent Hodge was locked up with the pirate Conyngham. Plenty of other schemes and intrigues were involving the various governments at this time and everybody forgot about Conyngham, allowing the ever-patient Franklin to arrange for the release of Conyngham and his crew.

This time **Revenge** set sail for the approaches to London where he scoured the shipping lanes for quarry. After a successful week, Conyngham sailed west, rounded Scotland's Shetland Islands and patrolled off Ireland's west coast but had difficulty finding British vessels that could be easily cowed. Conyngham headed the *Revenge* south and nearing the Spanish coast, encountered a British warship. The ships exchanged several ineffective shots, but *Revenge* having previously sustained storm damage affecting her maneuverability was vulnerable. Conyngham broke off the fight and entered El Ferrol, on the northwestern coast of Spain. Learning of Conyngham's skirmish with a British warship it was assumed that he was taking refuge in Spain, so they now sent a diplomatic protest to the Spanish government.

Conyngham had a new masthead constructed and other repairs attended to, all supervised by local American agents. Concerning payment he stated in a document: "*I pledge in case of need, my person, my belongings present and future, and generally and especially the armed sloop-of-war the Revenge which I command and the prizes and ransoms that I have already taken in virtue of a commission from the Congress. . . . .* etc," Conyngham still believed that he was in command of a Continental Navy ship and even took personal responsibility for paying the ships bills.

On receiving the British protest, and in order to ensure Spain's neutrality status, the governor of El Ferrol ordered Conyngham to depart. Soon

Revenge was cruising east of Bilbao, where she stopped a French brig, the *Graciosa* bound from London to La Coruña with dry goods fully insured in England. British merchants were now shipping their goods via neutral ships to avoid capture by the notorious pirates aboard *Revenge.* Taking a captured neutral French vessel into a neutral Spanish port was a huge diplomatic error. Authorities in St. Sebastián, arrested and jailed Conyngham's men. The captured brig *Graciosa* was rapidly returned to her owners.

When the details of the incident reached the American commissioners in Paris, they were for once in agreement, both being aghast by Conyngham's lack of judgment. He was considered diplomatically retarded. Berated, Conyngham, questioned "If British naval ships could confiscate goods from American vessels, then Continental Navy vessels should be able to do the same. Had they not a right to retaliate?" Conyngham clearly presumed he was acting under his Continental Navy captain's commission and had the prerogative to engage the enemy. He had no inkling that he might have been misled concerning the ships registry in the Continental Navy.

Conyngham continued harrying British shipping, now sending the prizes trans-Atlantic to

Newburyport, Massachusetts. On February 6, 1778, the Treaty of Alliance that recognized the United States of America was signed by France and a month after, war was declared on Great Britain. French ports were now a safe haven for Conyngham, but he continued to patrol off neutral Spanish territory because news had not reached him. When Conyngham put into Cadiz, Spain March 25, 1778 he heard the news. After repairs and replenishing stores, he slipped out to sea during the night, evading a pair of frigates sent to hunt him and laid course for the Canary Islands. While continuing to evade other frigates sent to hunt him, he captured and burned more ships including the Swedish brig *Henerica Sophia* deemed an illegal seizure. Only neutral vessels loaded with *warlike stores* and bound to the *Ports of the Enemy* could be lawfully detained. Carrying a cargo of dry goods, and bound for the Canary Island port, Teneriffe, she was exempt. Letters of rebuke about Conyngham arrived from assorted foreign diplomats.

Sailing for the Caribbean, *Revenge* took several ships, then made port in Martinique. Receiving a dispatch that Admiral Comte d'Estaing was bringing a French fleet and troops to Martinique, there was alarm because a squadron of British vessels were waiting in the area. On December 28 the Continental agent sent Conyngham to warn d'Estaing that a British squadron was likely downwind of the French fleet. Forewarned by Conyngham and aware of the British vulnerability, d' Estaing surprised and dispersed the British force.

Revenge returned to patrolling and captured or sunk over sixty ships, in the next year and a half. Insurance rates soared. Arriving back in the British Isles his ship suffered storm damage. Bold as brass, Conyngham disguising the *Revenge* and entered a British port for repair, all the while speaking his native Irish tongue to avoid suspicion. His campaign of destruction off the British coast only ceased when *Revenge* captured a ship whose cargo was war-material. Conyngham recognized its extreme value to Continental forces and decided to escort the prize to an American port. Making port in Philadelphia on February 21, 1779 with a treasure in military equipment, he should have been well received. Not so.

It is sad that the officials of the first theoretically friendly port Conyngham had seen in three years would treat him so shabbily. Though the newspapers greeted him as a returning war hero, the partisan Continental Congress made his existence a misery. Members were displeased with his disobedience to orders and the confiscation by the French of his original commission. Conyngham was astounded by the unrelenting hail of criticism and complaints thrown at him. The Marine Committee of Congress demanded *"that Capt. Conyngham give an account of himself."*

Conyngham found he was being tied to the scandalous misdeeds of some of the Commissioners he had dealt with in France. Former crewmen, whose return preceded his claim that they hadn't received shares or wages They alleged that both *Surprise* and *Revenge* were privateers and several partners along with Deane had profited handsomely from the prize money derived from Conyngham's cruises.

Of course, Conyngham's initial activity had occurred when France was trying to maintain political neutrality. The operations were disguised, his prizes and actions were approved and encouraged by American commissioners. Subsequently, even if irregular, the maritime operations proved to be an effective strategy during the early stages of the Revolutionary War. Conyngham claimed his only motive had been to harass British shipping in their dangerous home waters. Any resultant monetary gains were inferior to his goal. Conyngham's arguments gained sympathy, if not complete success in all quarters. This did nothing to contain the ire of some politicians.

Conyngham was no longer captain in the Continental Navy or of what he had assumed was the *"Continental Navy Cutter Revenge."* Dismissed, he formed the merchant firm of *Conyngham and Nesbitt* which purchased *Revenge* and Gustavus Conyngham became her civilian captain. For the period of a week, he chartered *Revenge* as a commerce patrol vessel to the State of Pennsylvania. She was to protect the city's river under a letter of marque and as a lawful privateer. Off the Delaware Capes on a private cruise, the Letter of Marque expired. Unable to outrun the British warship *HMS Galatea*, *Revenge* was boarded and able to produce only what were expired papers, Conyngham was taken aboard as a prisoner. *Galatea* docked in Tory, New York, where Conyngham was put aboard a prison hulk anchored off Brooklyn, then weighted down with fifty-five pounds of chains fastened to his ankles, wrists and about his neck and kept at the provost's prison.

Eventually brought before Commodore Sir George Collier, he was ordered extradited to Britain to be hanged. Conyngham was first paraded through the streets on a cart, before being put on the

British Packet *Sandwich* for London. Having caused bitter embarrassment to the British navy over a three-year period, there was a great deal of personal animosity. Not properly fed, he lost fifty pounds while being transported to a British prison. Charged with piracy and incarcerated at Pendennis Castle, he promptly escaped but was caught and put in Mill Prison at Plymouth. When they announced he was to be sent to the gallows, General George Washington, informed the British that if they hung Conyngham, he would reciprocate with six British officers in his possession, a threat that cooled things down.

Conyngham, then was tried for treason but before sentencing, he escaped again, this time with eleven other prisoners, by digging a tunnel underneath the outer wall of the prison. Managing to cross the channel to Netherlands, he planned to find a way back to America where potentially he could receive a new ship. He was in the port of Texel when John Paul Jones entered after his battle with *HMS Serapis*. Joining Jones on his new frigate, the *Alliance*, he sailed with Jones for over three months.

Returning to America from Spain as a passenger, Conyngham was captured again when the *Experiment* was caught by the British navy. He was imprisoned a third time at Mill Prison but was soon released in a prisoner exchange. Going to Ostend, he was fitting out the armed ship *Layona* when news of peace reached him. Gaining passage on the *Hannibal*, he was soon bound for his home in Philadelphia.

Back home he asked for the years of back pay due him. Conyngham's captain's rank in the Continental Navy was denied and they refused to pay him for his naval service. He had been given two naval captain's commissions, signed by the President of Congress and one personally presented by the esteemed Benjamin Franklin, and letters of Marque but Congress now ruled these commissions were considered temporary or non-existent. Alexander Hamilton, the Treasury Secretary tried to see he received justice but failed.

Enduring many hardships Gustavus Conyngham served the American cause with distinction. Conyngham destroyed dozens of enemy vessels and took many small ones and sent thirty-one major prizes to port, more than any other American naval officer in the Revolutionary War. He received almost nothing. Having been issued a Continental Navy commission, he presumed his actions to be within the rules of maritime war but due to his government's agents, no official record of *Surprise* or *Revenge* existed on the registry of the Continental Navy or as Letter of Marque vessels. Congress didn't care, and legally Conyngham could have been tried, convicted, and hanged as an unknowing pirate. He could have used a good attorney. **Gustavus Conyngham** died in Philadelphia on November 27, 1819 at the advanced age of seventy-two, having contributed greatly to the realization of America's independence, he was never valued for his sacrifice.

The War of 1812

Privateer Brig Yankee 1813

By the War of 1812 the United States had a far better system of government. The navy had been neglected but the framework was there and there were some good ships, but not enough. The call for privateers went up as before. President Madison issued 500 Letters of Marque, authorizing privateers. Overall, some 200 of the ships took prizes. The cost of buying and fitting of a large privateer was about $40,000 and prizes could net $100,000 and more.

The single most successful owner, operator of privateers in the War of 1812 was **JAMES D'WOLF** US Senator from Bristol, Rhode Island. Below is his letter to the Sec of War requesting paperwork for his first privateering vessel, the *Yankee*.

BBISTOL, R, I,

June 30, 1812,

The Honorable WILLIAM EUSTIS, Secretary of War:

Sir; I have purchased and now ready for sea, an armed brig, one of the most suitable in this country for a privateer, of one hundred and sixty tons burden, mounting eighteen guns, and carries one hundred and twenty men, called, the Yankee, commanded by Oliver Wilson. Being desirous that she should be on her cruise as soon as possible, I beg that you will cause a commission to be forwarded as soon as practicable to the Collector of the District, that this vessel may not be detained. I am very respectfully. Sir, Your obedient servant, JAMES DE WOLF.

On July 13 of 1812, the Yankee's commission was allotted, listing her owners as: James DeWolf, owning 75% and John Smith, owning 25% of the vessel. Officers: Oliver Wilson (aged 26 at the time), was Captain; Manly Sweet, James Usher, and Thomas H. Russell, Lieutenants. The Articles of Agreement under which the privateer sailed were as follows:

ARTICLES OF AGREEMENT BETWEEN THE OWNERS, OFFICERS AND COMPANY OF THE PRIVATE ARMED VESSEL OF WAR, "YANKEE."

1st, It is agreed by the parties that the Owners fit the Vessel for sea and provide her with great guns, small arms, powder, shot and all other warlike stores, also with suitable medicines and every other thing necessary for such a vessel and her cruise for all of which no deduction is to be made from the shares, for which the Owners or their substitutes shall receive or draw One Half the net proceeds of all such Prizes or prize as may be taken, and the other half shall be the property of the Vessel's Company to be divided in proportions as mentioned in the 15th article, except the cabin-stores and furniture which belong to the Captain.

2nd. That for preserving due decorum on board said vessel, no man is to quit or go out of her on board any other vessel, or on shore without having first obtained leave of the Commanding officer on board, under the penalty of such punishment or fine as shall be decreed by the Captain and Officers. 3d. That the Cruise shall be where the Owners or the major part of them shall direct.

4th. If any person shall be found a RINGLEADER of any Mutiny, or causing disturbance, or refuse to obey the Captain, or any Officer, behave with Cowardice, or get drunk in time of action, he or they shall forfeit his or their shares of any dividend, or be otherwise punished at the discretion of the Captain and Officers.

5th. If any person shall steal or convert to his own use any part of a prize or prizes, or be found pilfering any money or other things belonging to this Vessel, her Officers, or Company, and be thereof convicted by her Officers, he shall be punished and forfeit as aforesaid. '

6th. That whoever first spies a prize or sail, that proves worth 100 dollars a share, shall receive Fifty Dollars from the gross sum; and if orders are given for boarding, the first man on the deck of the

Enemy shall receive Half a share to be deducted from the gross sum of prize-money.

7th. That if any one of the said Company shall in time of action lose an eye or a joint, he shall receive Fifty Dollars, and if he lose a leg or an arm, he shall receive Three Hundred Dollars to be deducted out of the Gross sum of Prize-money.

8th. That if any of said Company shall strike or assault any male prisoner, or rudely treat any female prisoner, he shall be punished or fined as the Officers shall decree.

9th. That if any of the said Company shall die or be killed in the voyage, and any prizes be taken before or during the action in which he is so killed, his share or shares shall be paid to his legal representatives.

10th. That whoever deserts the said Vessel, within the time hereinafter mentioned, shall forfeit his Prize-money to the Owners and Company of the said Vessel, his debts to any person on board being first paid out of it, provided it does not amount to more than one half the same.

11th. That on the death of the Captain, the command to devolve on the next in command and so in rotation.

12th. *That no one of said company shall sell any more than one half his share or right of claim thereto of any prize previous to her being taken.*

13th. *That the Captain and Officers shall appoint an agent of said Vessel's company for and during the term of the said cruise.*

14th. *That all and everyone of said Company do agree to serve on board of said Vessel for the term of four months, conformable to the terms herein mentioned, beginning the. said term at the time of her departure from the harbor of Bristol.*

15th. *That One Half of the net proceeds of all prizes taken by the said Vessel which is appropriated to the Vessel's Company shall be divided among them in the following manner (viz)*

To the Captain sixteen Shares and all such privileges and freedoms as are allowed to the Captains of Private armed Vessels of War from this Port. To the First Lieutenant nine Shares.

To the 2d and 3d Lieutenants and Surgeon eight Shares each.

Prize masters and Master's Mate and Captain of Marines six Shares each.

Carpenter, Boatswain and Gunner four Shares each.

Boatswain's Mates two- and one-half Shares each.

The residue to be divided among the Company in equal Shares excepting Landsmen or raw hands who draw one- and one-half shares each, and boys who draw one Share each. Ten Shares to be reserved to the order of the Captain to be distributed by him to such as he may deem deserving among the Vessel's Company."

D'Wolf's, *Yankee* brig was enormously successful from the moment she cast off, and this was not a common situation among privateers of the "War of 1812." Contrary to common belief, privateering was a financially risky business. Only a small percentage broke even financially and many lost their entire investment. The very threat of privateers was a massive drain on the enemy nations shipping profits, though. Having an overly fast ship with excellent sailing ability's and a crew that knew how to use them was of paramount importance but so was dumb luck. The *Yankee* and her crew of one-hundred-fifteen had all the qualities required. On the first cruise toward Nova Scotia, *Yankee* took ten prizes, one in a running fight with a vessel of three times her tonnage.

The *Yankee* made numerous successful forays during the war, possibly being the most financially successful privateer in the war but there was a cost for crew members. Shortly before his ship returned home, one Captain ended a letter to his wife as follows:

"P. S.-- I have lost one of my legs on this cruise."

Exerts from a privateer's log:

Life at sea was not all calm water, sunshine and pleasant breezes.

9th Day: Fair weather with strong gales from the westward. Scudding before the wind under square-foresail, fore topsail and foretopmast staysail. At 5 pm, we discovered from the deck (owing to the negligence of his duty the man at the foretop) sails of two large ships following in our wake, distance about three leagues, standing after us with their topgallant sails up. Immediately we hauled up to the S. E. and set square-foresail, single-reefed mainsail and fore and aft foresail. The sails astern, unable to follow us up to wind, observed frequently luffed up and yawned off. When we saw them last, they had stood off to the N. E. Frequent squalls with rain and a tremendous sea. Course S, E, by E. under three-reefed mainsail close-reefed square-foresail, and double-reefed foretopsail, with the foretopmast-staysail. Same persons on the Surgeon's list. Shipped a great deal of water upon deck, the coamings of the sea frequently coming on board and penetrating every part of the vessel. Lat. 34°40'.

10th Day: No sail in sight and nothing remarkable, Lat. Obs. 33°26'. N. B. It is something singular that since we left port, we have had only one pleasant day. There has been a continual succession of gales of wind from all parts of the compass, attended with torrents of rain, squalls, whirlwinds, thunder and lightning, and a tremendous sea frequently breaking on board and occasioning considerable damage; to include carrying away several spars and staving the arm-chests. Indeed, it may be said that our vessel has sailed thus far under----not over the Atlantic Ocean.

11th Day: Middle and latter part of the day stiff gales with a high sea. Shipped a great deal of water upon deck. Lat, Obs. 32°5'.

12th Day: During these 24 hours strong gales with frequent squalls of wind and rain, and a very high sea frequently breaking on board, Lat, Obs. 30°27'. Lunar Observation at 23 M. past Meridian 41°55'41". Cyrus Simmons, John Briggs, Amos A. Allen, James Angelí, Ebenezer Byrum and William Redding unfit and on the Surgeon's list.

A chase and violent action between ships and crews did not end with the last shot. Crews interacted after a capture.

46th Day: At 7 a. m. Abner Midget spied a sail right ahead distant about 5 leagues. Got out all the sweeps. 8 a m observed several waterspouts under the lee—squally with flying clouds and rain. At 11 made out the chase to be a schooner standing to the eastward. At meridian still in chase of the schooner distant about 2 leagues. Lat. Observation. 6°55'.

47th Day: At noon meridian continued in chase of the sail ahead as before moving the ship with the use of sweeps. Just past 12 launched our boats to assist the sweeps by towing. This has brought us up rapidly with the chase. 2 p m, we fired a gun; hoisted English colors; not answered. Just past 2 p m, gave her a gun, upon which the chase showed English colors. 3 p m, being distance of about 1 mile we hoisted American colors and commended firing Long Tom, while towing the Brig all the time with the boats. 4 p m, got the boats astern, piped all hands to Quarters and cleared for action. Now moving in light airs and a smooth sea. Being now within good, gunshot commenced a brisk cannonade on the starboard side. The chase returned the fire with 4 guns, the shot frequently falling near and one shot wounding the jib. At 20 minutes past 4 p m, the Enemy fired a stern-chaser, double-charged. It instantly blew up, occasioning a tremendous explosion.

Observed the schooner to be on fire and several men swimming alongside. We immediately ceased firing (although her colors were still flying). We sent our boats with Lieut, Barton and Master Snow on board to save the lives of the enemy and extinguish the fire. They took up the swimmers and then rowed alongside. The scene that now presented itself to their view was shocking beyond description. The vessel was still in flames, the quarter-deck was blown off, the Captain was found near the mainmast—naked, mangled and burnt in the most shocking manner. One of the seamen lay near bruised and burnt almost as bad, a black man was found dead on the cabin floor, and five others around him apparently dying. All these wounded men were sent on board the Yankee and there received every possible attention from the Captain, Surgeon and Officers. Doctor Miller dressed their wounds and gave them the proper medicines but found the Adler's Captain and several of the blacks in a most dangerous condition. The Captain had received two deep wounds in the head which penetrated to the skull (probably from our langrage shot), his arms and legs were much bruised, his skin nearly all burnt off and his whole system greatly injured by the concussion. A small black boy had a most singular yet distressing appearance. This boy was literally blown out of his skin and for some time after he came on-board, we thought he was a white child. The sufferings of these poor fellows seemed very painful and excruciating.

Lieut. Barton extinguished the fire and sent all the prisoners on board together with a boatload of sundry articles taken out of the cabin which had not been consumed by fire. Finding the prize, no ways injured except in her quarter deck the Commander ordered Lieut, Barton with a chosen crew to remain on board and to keep company with us during the night. On examination of the Schooner's papers and logbook we found her to be the Letter of Marque Schooner called The Alder of Liverpool, (owned by Charles B.

Whitehead) formerly called La Clarisse and taken from the French. She was commanded by Edward Crowley, 77 tons burthen, mounting 4 carriage guns, and navigated by 10 men, besides 11 African crew, men. She left Sierra Leone 9 days past, bound to the Leeward on a trading voyage, with an assorted cargo of bafta (Indian Cotton),' gunpowder, muskets, bar-lead & iron, beads, flints and sundries. The Alder appears to be about 4 years old, is copper-bottomed, measures 67 feet in length, but her sails are very poor, and she does not sail well. The probable value of this prize in America might be $5000; but her net value could not exceed $3000.

At 8 p. m. one of the black seamen died and was thrown overboard. 25 minutes past 2 a m, Captain Crowley notwithstanding every medical assistance departed this life in the greatest agony. For some hours previous to his dissolution he appeared to suffer excruciating torments and when informed of his approaching demise did not seem sensible of his situation. His body was committed to the waves with as much decency as was practicable. At 9 a m, the boy before mentioned also died and had a watery grave. The white seamen and three other blacks are just wavering between life and death and we fear they cannot recover.

The schooner's boatswain related to us the accident which led to the horrid catastrophe. He said the Captain stood at the helm steering the vessel and giving his orders; and that he himself and several of the seamen were stationed at the gun aft; that the instant it was discharged the gun burst with great violence, and broke one of the quarter deck planks. It threw the wood-all on fire—directly into the powder magazine which was situated abaft the cabin, and the vessel instantly blew up. (Himself and another seaman leaped into the sea when they saw the gun dismounted and thus saved themselves.) It is supposed the Captain was thrown from the helm, up into the air and then landed in the main rigging. The blacks who were so dreadfully mangled were in the magazine filling cartridges.

I had sent the carpenter with materials to repair the prize, which is well for at 4 a m, there came upon us one of the most tremendous squalls and tornados ever witnessed. It blew, rained, thundered and lightninged in a truly terrific manner. We took in all sail and kept the vessel scudding before it. The lightning was unusually vivid and struck several times close on board. Having no conductor every mind was filled with apprehension and alarm. Latter it fell to very light airs inclining to calm.

48th Day: During the greater part of these 24 hours calm with occasional light airs. At 4 and 6 p. m. the two other black seamen who were blown up on board the schooner died and thus were thrown overboard, making altogether six persons who have perished by this most unfortunate accident. The white seaman is still in a most dangerous state, but the Surgeon gives us hopes of his recovery. We were much surprised on examination of the Alder's flag box to discover a Pirate's flag and pendant. This circumstance lessens our compassion for the deceased Captain Crowley as it indicates an evil and hostile disposition toward all mankind.

At 5 p m, we came to anchor on the north side of the Island of Annabona in 7 fathoms water, sandy bottom, opposite a small village distant about a mile from the shore. Soon after we came in, the black Governor and his mate came on board. We easily obtained permission to water, wood etc. Having finished our trade and filled our water at 20 m before midnight, we got under way for sea.

Vices held by men ashore cannot be allowed at sea. The captain wrote.

It is with deep regret that the Commander of the Yankee feels it his duty, in justice to himself, his Officers and his crew, to make the following entry in the ships Journal, relative to the ' conduct of one of his officers.

*My Second Lieutenant John H. Vinson, has never, in my opinion, displayed either seamanship, judgment or courage during our cruise. He appeared to be much intoxicated on the night of the partial engagement with His Majesty's Schooner Saint Jago and behaved with great impropriety. During the skirmish with the Alder he was particularly negligent in not extinguishing the flames when our own bulwarks were on fire. And during the long engagement with the Andalusia he certainly did not manifest either activity or courage. This officer is guilty of one grievous offence which would subject him even to capital punishment: that of sleeping on his watch. The night after we captured the Fly, when we had a number of prisoners on board, and many of our crew had got drunk on board the prize, and were extremely riotous, Lieut. Vinson was himself much intoxicated, or to speak plainly, **dead drunk**, and slept on his watch in presence of myself, my officers, and the whole crew. He was guilty of the same offence on the 5th January when we lay at the Gaboon, and also last night when we had 14 prisoners on board that were disposed to rise up and cause us injury at any time as we were anchored on a savage coast. This offence of* *getting drunk and sleeping on a watch is of a very serious and alarming nature, endangering both the safety of the vessel and the lives of all on board. His conduct subjects him to a court martial which will certainly convince him of his errors.*

Such were day to day activities aboard a privateer.

In the less than three years that the *Yankee* cruised the seas as a Privately Armed Vessel of War, she captured more than five million dollars of British property. She brought back to the town Bristol crew shares of more than a million dollars as the profit from' her six cruises, a tiny fraction of what it is inflated to today. No other American Privateer sailing from an American port ever established such a record.

Thomas Boyle's Chasseur

A flamboyant and well-known Privateer, **Thomas Boyle** was another of the very successful privateers of the War of 1812. The ***Comet*** was his first ship and with her he once captured a ship that could have hoisted the *Comet* aboard like a barge. The success of the first ship funded his second ship, a ship whose speed became a legend. His ***Chasseur*** was so popular in the Chesapeake it was nicknamed the Pride of Baltimore. The *Chasseur* mounted 16 twelve-pound guns, and 2 carronades. Captain Boyle captured 18 rich British merchant ships on his first voyage with *Chasseur*. He so irritated one British admiral that a squadron of fast naval vessels was tasked to hunt him down. Boyle's *Chasseur* boxed by two frigates and a pair of sloops out maneuvered and out sailed them. Later, two navy brigs, set a trap. Hoisting the Stars and Stripes, Boyle sailed straight up toward one of them, tacked and firing a shot, hauled to windward an angle the Brig couldn't dream of following. In the several days that followed, *Chasseur* out sailed flotillas of three, four, and five ships, and did it with ease.

Once, late in the war, he was encircled in light winds. Boyle jettisoned ten of his 12-pound cannon, along with stores and spares sails. Guns were shifted aft and the stern rails were cut away to give them a wide arch of fire. Each time a gun was fired at a pursuing enemy ship, Chasseur lurched ahead 6 feet. Soon, she picked up a breeze and slipped away.

At the end of the war, Boyle encountered a large schooner. As *Chasseur* closed, they saw only three gun-ports to a side and what appeared to be a small crew. As *Chasseur* was brought to hailing range, many men rose from hiding and *HMS St. Lawrence* fired a broadside from 10 hidden guns. Unable to catch *Chasseur* they had sent a decoy ship. *Chasseur's* deadly return fire chewed the navy schooner up in a fifteen-minute fight though. *HMS St Lawrence* surrendered; on this occasion a privateer had captured a Royal Navy ship.

Privateer America

The privateer *America* was another vessel that frustrated British efforts to put her out of business. Having a well-deserved reputation for plundering ships and cargos the *America* went on the admiralty's most wanted list. To end the devastating marauding the British government built a fast frigate, the *Dublin*, for the express purpose of chasing the *America* down. The *Dublin's* captain seemed always to be staring at *America's* stern.

Long after the war had ended, the captain of *America* and the captain of the *Dublin* met in Valparaiso. Neither man was aware of the other's history. During a pleasant conversation that had drifted to the subject of the war, the Briton remarked:

"I was once almost within gun-shot of that infernal Yankee skimming-dish. It was just as night was coming on. By daylight she had out sailed my Dublin so devilish fast that she was no more than a speck on the horizon. By the way, I wonder if you happen to know? What was the name of the beggar that was master of her?"

"I'm the beggar," smiled the American master and they drank a toast to each other's health.

The privateer that won the battle of New Orleans.

The Portuguese Azores Islands are about one third of the way across the Atlantic from Africa and have been a watering and provisioning stop for centuries. In 1814 Portugal remained neutral in the conflict between the United States and the British Empire. At dusk of the 26th of September 1814 the American privateer, *General Armstrong* was taking on water and supplies in the port of Fayal. A British squadron en-route with troops for the assault on New Orleans had arrived off Fayal and on reports of an American warship, three men-of-war approached the harbor. With orders to sail with dispatch to join the flotilla assembling in the Gulf of Mexico with dispatch, the British commander's decision to capture a Privateer inside a neutral port seem a little crazy but that's what he did.

Boats were first launched to scout but fell back quickly when bracketed by canon fire.

Barges were then launched from *HMS Plantagenet*, and *HMS Rota* at about 8 pm. By 11 pm 180 marines boarded the boats, at midnight, under cover of dark they attack: attempting to board by the schooner's starboard aft quarter, as well as at the bow. The Americans opened fire with long 9-pounders and a swivel gun at the rail, while the barges crew fired back with small carronades. Suspicious of British intentions, Captain Reid of the *General Armstrong* had ordered boarding nets hung after dark. On the quarter these nets thwarted the British plan to move quickly and overcome the schooner's crew. Unable to clear the netting on the quarter, pistols and muskets were being discharged at them at point blank range; long pikes were thrust in their faces. Bloodied and reduced they dropped back down into their boats. Meanwhile the attack over the bow was a more equal contest and could have succeeded had not the captain rushed forward with reserves from astern.

On the following day the crew of *General Armstrong* was not surprised to see *HMS Carnation* entering the harbor attack the schooner, but she bore off when the privateer's big 42-pound carronades began to damage her rigging. Though successful so far, Captain Reid was a realistic man and quite practical. Long term chances of retaining his ship were short to nil. So far, he had only 2 men killed. He quietly began moving his crew and some valuable cargo ashore. When he noticed renewed activity among the British, he fired a canon down the hatch, blowing a large hole in the *General Armstrong's* bottom. Rowing ashore to join his men, Reid observed the British preparing to assault his sinking ship.

This action had cost the British squadron inordinately, up to 250 casualties. Delayed by reduced manpower and time needed to deal with wounded, the squadron was delayed by nearly two weeks, and they arrived at New Orleans late. The squadron's delay forced Major General Sir Edward Pakenham to postpone his advance on the city, giving General Andrew Jackson time to arrive and organize the city's defenses.

Meeting Reid socially, years later, General Andrew Jackson was reported to have told him, *"If there had been no Battle of Fayal, there would have been no victorious Battle of New Orleans for the US."* Reid had delayed the British expedition against New Orleans for ten days allowing Jackson to arrive there before them. Thus, Louisiana and the Northwest Territory might now be British if Reid had not engaged them in what has been called one of the world's most decisive naval battles. The Lower Mississippi might now be British.

Of course, there were also British, French, Dutch, Spanish and other privateers of the Napoleonic period that were very good at what they did and more Americans than I have paper for.

The 1856 Declaration of Paris, outlawed privateering. The US was not one of the initial signatories. When the Confederate Congress authorized use of privateers the Federal government signed the declaration right quick. Thus, some of the world's last privateers sailed under the flag of the Confederate States of America.

The Confederate Raider Alabama sinking the USNS Hatteras.

Postal Packet Vessels

News, messages, intelligence, and in the last few centuries regular mail service has been important to civilizations. On land, roads and routes were established, to move armies, carry commerce, and to dispatch important messages. Communication was necessary to hold empires together and runners and riders have dashed from one land to another with important communication for several millennia. When the roads reached the sea though, the messenger needed passage over it.

Before even the Sumerians, Minoans, Mycenae, Phoenicians and Greeks dispatches needed to cross over the world's waters. Some of the first vessels chose for the task would have been log canoes, then there were light galleys, (Moneres) or small vessels with a single bank of oars, rowed by as few as 20 men. Sail assisted galleys of this type were given different names like Greek Penteconter, the Roman Cisocontores. They were fast rowing vessels, able to enter shallow rivers or to be beached if need be.

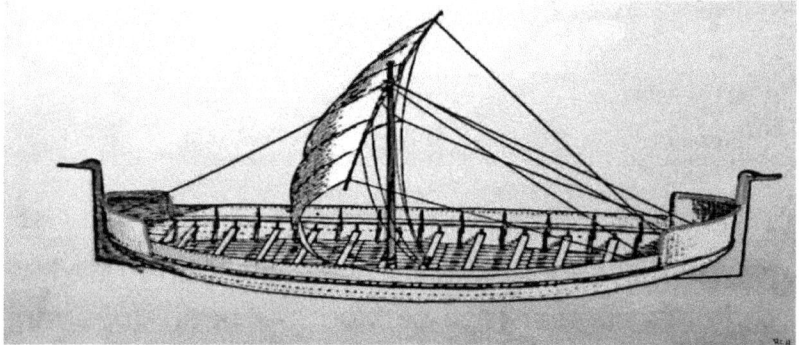

Cisocontores

The Romans, not by nature a sea going people, first adapted a vessel common to the south of Rome and on the island of Sicily, the Penteconter. This type of galley had been around since before the Trojan War, in use before 1400 BC. The

Penteconter was "Romanized" not as a military vessel but as a boat that could deliver dispatches, liaison, carry a messenger or envoy, even scout a far shore. Different types of the vessels which later came to be known as Packets, were developed all over the world and their development and design reflected the waters they swam in. In many locals they remained powered by oars; in others they were driven primarily by sail.

In Western Europe, the use of lightly armed fast vessels to carry messages, and diplomats began in the Tudor period. Dispatch boats, later known as packets for parcels or packages, have been widely used for coastal mail services since. By the end of the 16th century, with longer voyages, some minimal accommodation was made for passengers. A person taking passage on a packet to the Americas or even to the distant Far East, made do with water and a box to sleep in. Often you were expected to bring your own food and use the galley fire to cook your own meal. In the 1600s up to twenty royal packets were put in service by the English monarchy to maintain communications with Europe and Ireland. Others were eventually commissioned to maintain communications with the American Colonies and a century later in India and Australia.

Coffee houses and taverns were known as public houses and were the information centers of ports and cities. Mail and packages were usually gathered at a public or common location, often a room or warehouse adjacent a tavern. On shore, a ship's captains might use such locations as a place to conduct business. Passengers could room while waiting to board ship, hear the latest news. Anyone could drop a letter in the ship's mail bag and pay the captain's representative for delivery.

In 1710, the British parliament established an important law, funding packets. New packets were all to be ocean going ships designed for speed. Some passengers and cargo could be carried but the packet's primary purpose was to carry mail and small parcels. Secondary was transporting officials or military personnel to their stations and to do this on a regulated schedule. It was forbidden to lay over or delay waiting to top off the hold, rather it was to sail as the timetable determined with a single letter if that's all that had come aboard.

Various national Postal Services used Packet Ships to carry mail packets to and from embassies, fleets, colonies and outposts. The vessels might also carry the crew's private cargos as well as coin or bullion. In western Europe, the rigs and sizes varied widely: sloops, lugers and cutters were common early on but the need for size caused the evolution to schooners, brigs and brigantines. In the Mediterranean galleys, feluccas, and xebecs were more common. Bounties were offered for fast passages and a record passage could make a captain famous.

Speed also served to protect the ship from being taken by enemy naval vessels or privateers in time of war or pirates anytime. With minimal armament, packets relied on speed and for centuries with the world constantly at war, they met their match occasionally in a fast pirate or privateer. In the resulting navel engagements, there were many incidents recorded of both successful defense and of capture. Often the same ship might be captured and bought back a number of times over a long career. Dependent on circumstances a packet could come out victorious. During the Napoleonic War, there were many such incidents. One, in which the 6-gun Packet, *Princess Royal*, bound toward New York, was overhauled by a French privateer, also a brig but carrying 16 guns and nearly twice the crew. For a period of several hours, the two ships engaged in a running battle, during which passengers and even children aboard the packet joined in the defense. Finally, the French privateer sustained so much damage that she was forced to break off her attack and return to port for repairs.

Postal Packet Sloop

Ships were often captured and retaken in a matter of days. One such occasion was when the French 18-gun privateer brig, *Insolent*, captured the

Walsingham Packet, which was bound for Lisbon. She was retaken 12 days later, by the British naval vessels, *HMS Porcupine,* a 6th rate of 26 guns and the *HMS Minotaur* a 3rd rate 2 decker. They caught the prize crew before they could get her into a French port. In fact, the privateer *Insolent* barely escaped into the port of Lorient, herself.

The case of the packet, *Marquis of Kildare* was even more back and forth. In 1800, the French captured the packet *Jane*, and the Captain and many of her crew were put ashore in the Portuguese port of Lisbon. Wishing to return to England, they booked passage on the Lisbon packet *Marquis of Kildare*, bound for Falmouth. Before the *Marquis of Kildare* reached port, she was captured by another French privateer which took her captain, officers and crew off and sent the packet to France with a prize crew. The Captain of the *Jane,* some of his crew, and a few other passengers however had managed to hide in the hold. In the dark of night, they retook the ship, and put the French prize crew over the side in a long boat. With only 7 men to handle her, and great difficulty, the *Marquis of Kildare* managed to make a port in Cornwall two weeks later.

A less successful meeting took place when the ship-rigged packet, *Princess Amelia* was overhauled by the American fast-schooner, *Rossie*. This brief but hot battle took place during the war of 1812 and ended with eleven of *Princess Amelia's* crew wounded, her captain and two others dead. The *Princess Amelia* struck her colors within minutes. Against twice the number of heavier cannon and three times the crew, the packet stood little chance against the privateer *Rossie*.

This is not to infer that a lightly armed packet couldn't come out on top in a fight. In the case of *HM Packet Antelope,* for instance, the packet fought it out with the French privateer *Atlante* in 1793. With half the privateer's guns and crew, the packet actually defeated and captured the privateer though the packet's captain and officers were killed in the action. One of the most celebrated, against all odds, engagements took place in 1807 in the Eastern Caribbean. The packet *Windsor Castle*, a brig crewed by 28 men and armed with 8 guns was approaching her destination of Barbados with a cargo of dispatches and packages. The French privateer, *Jeune Richard* was patrolling a hundred miles North East of the island, in hopes of snatching up shipping arriving from Europe.

Not long after dawn, the lookout at the *Windsor Castle's* masthead reported a sail approaching from astern. After watching her for a quarter hour it became obvious; clearly, she was very quick. The *Windsor Castle* was considered a fast ship and so the speed with which the unknown vessel was overtaking was disconcerting to her captain and crew. The rigging had now topped the horizon enough to identify it as a schooner but there was no hint yet of its nationality.

All additional sail, including stunt sails were set and as wartime policy directed, Captain William Rogers ordered the bags of official mail to be weighted and secured at the rail so it could be jettisoned and kept from enemy hands in the event the ship was taken. By eleven it was evident that they would be overtaken. Rogers ordered the crew to quarters and began to prepare the ship for a fight. Weapons were broken out and boarding nets secured on both port and starboard rails.

At 12:15 *Jeune Richard* hoisted the French Tricolor and fired her bow chasers. *Windsor Castle* responded with guns on the stern. Not strikes, but now both sides knew of the others' intention to fight. Their nations were at war, but a privateer is a business enterprise and the ship, and its crew are at sea to make money. There is no profit in sinking a vessel, rather they are out to capture it, to auction it and its cargo. This puts the privateer to a disadvantage it a fight. Their goal is to scare the enemy crew into surrender and at worst cripple the vessel and board it.

The canon fire had caused only minimal damage as the distance between the ships narrowed and by 13:00 she slammed alongside, grapnels flying over the rails and into the packet's rigging. The first French boarding party swung aboard but unable to get past the boarding nets were stopped. Some were skewered with pikes and the remainder leaped back aboard *Jeune Richard.* This was only a temporary relief for the British crew; they were clearly outnumbered three to one. The men on both ships kept up a steady small arms fire and the privateer crew made several weak attempts at boarding, all of which were repelled by the British.

It became plain to the *Jeune Richard's* Captain that only an all-out attack would succeed. Captain Rogers, aboard the *Windsor Castle* had come to the same conclusion. He ordered one of his ship's two carronades to be loaded with canister and for it to be positioned to fire at an angle along the Frenchie's deck. Shortly after 15:00 the French crew gathered in mass on deck to make another boarding attempt. Rogers's crew prepared to repel as before but when the French pressed forward to the rail the gunner fired the carronade. Loaded with hundreds of musket-balls the short-range carronade was similar to a giant shotgun and it swept across the packed Frenchmen like the grim-reaper's sickle.

Captain Rogers and a picked party boarded through the smoke of the horrendous blast. The remainder of the French crew were dazed, many of whom were wounded. Still numbering over fifty, they spirit was broken and infected with terror they fled below deck.

Recognizing there were still too many left to fight, and with time to collect themselves these men would regain their courage, Rogers negotiated a surrender and ordered them on deck, shackling one man at a time as they came up the ladder.

With the fighting over, it was evident that fortune had smiled on the crew of the *Windsor Castle*, for of her 28-man crew she had sustained only 3 dead and 8 wounded compared to 21 dead and 35 wounded out of the privateer's crew of 92.

It was understood that small ship actions were often the most vicious, battling locked together, men fighting tooth and nail, thus, the capture of the *Jeune Richard* was celebrated all over Britain. William Rogers and the crew of the *Windsor Castle* were well rewarded for their victory which was widely heralded in the papers and journals in Britain.

Over the centuries there were thousands of struggles as correspondence was passed over the earth's waters. In the 1820s Britain and other nations began switching postal service to steam vessels. In 1851 the last sailing packet was taken out of service and replaced with streamers. Within a decade most of the world's countries had done the same.

The Wreckers

Salvage schooner racing toward a stranded ship.

For over seven millennia men have ventured to sea fishing, exploring, trading, or for whatever reason suited them. For all this time they've understood it to be far from a safe endeavor, a calling leading to far more fatalities than professions on shore. Many a vessel, large and small has sailed out from land never to be seen again. Even so, it's not on the open ocean but at and near the coast where the greatest danger exists. Ships encountering reefs and banks or piling onto the shore itself were usually doomed as were those who sailed aboard them. For centuries the peoples inhabiting the coast looked upon wrecks as found treasure, wealth that they were fully entitled to. As if storms, fog, currents, mistakes in navigation weren't sufficient to put ships aground, on many of the world's coastlines populations did what they could to lure ships to their doom.

Men became wealthy by tricking them ashore and plundering them, a profession known as wrecking. Generally, crews and passengers drowned outright but if one was a rare swimmer, strong or lucky enough to reach the shore, he or she was most often murdered when they attempted to crawl from the water.

This story is not about those wreckers who purposely caused the destruction by luring ships to their doom. Rather, it's about salvors, known by the name wreckers, in that they worked with wrecks; these groups or crews of men whose profession it was to salvage a wreck. Their task was to save a ship or as much of its cargo as possible and be rewarded with a percentage of it's worth. These men were duty bound to save the passengers and crew as well. Most, if not all, followed this humane practice. It was often a blurry line separating the men involved in honest and criminal salvage. Maritime records are filled with incidents in which complete ships, with cargos, passengers and crews were saved by the expertise and courage of wreckers. Just as many reports exist, (mostly in earlier times) of what amounts to straightforward piracy. Practices such as the moving of navigation lights or as bad, the placing of false lights to cause the grounding, before acting the part of saviors. After removing cargo, the vessels would later be burned to remove evidence or more practical, so that the hull or rigging of the wreck wouldn't warn other ships that they were approaching a shoal. Although on many coasts men of both groups were called wreckers, the only similarity was that both groups removed cargo from wrecks.

One such industry evolved in the Florida Straits, a reef lined channel bordered by Florida, Cuba, and the Bahamas. It's a passage which extends from the Dry Tortugas, to a point north of the little Bahamas Bank, thru which the powerful Gulf Stream current flows. The straits are the outlet into the Atlantic from the western Caribbean and Gulf of Mexico and are bordered by banks and low islands from which coral reefs extended far offshore. These reefs often rise from great depth to mere feet in as little at a hundred yards. The first ships to come to grief on these reefs were those of the Spanish Treasure Fleets.

The Spanish fleets would be formed up in Havana's secure harbor and depart for Spain, sailing in convoy and carried north by the Gulfstream. In the mid-1500s the wrecks of Spanish treasure galleons were already scattered north along the dangerous reefs. Navigation was crude, lights were non-existent. Add blinding rain squalls, the occasional hurricanes and It was the perfect location for wreckers.

Floridian Indians, employed by the Spanish, became the first salvager divers in the area. Departing Havana in 1622, the Spanish Fleet encountered a hurricane within hours, and was driven ashore along the keys. Near the Marquesas, only 90 miles north of Havana, the *Atocha* and the *Margarita*, both major ships were wrecked in shallow water. Driven across sand flats by hurricane waves, the ships broke up slowly scattering treasure over the bottom for miles. The majority of what was recovered was accomplished over an eight-year period, by hiring Indian divers. Besides weather and remote locations, these operations had other problems. The salvage crews were regularly harassed and attacked by crews of English and Dutch ships who had hopes of taking home a little gold or silver themselves.

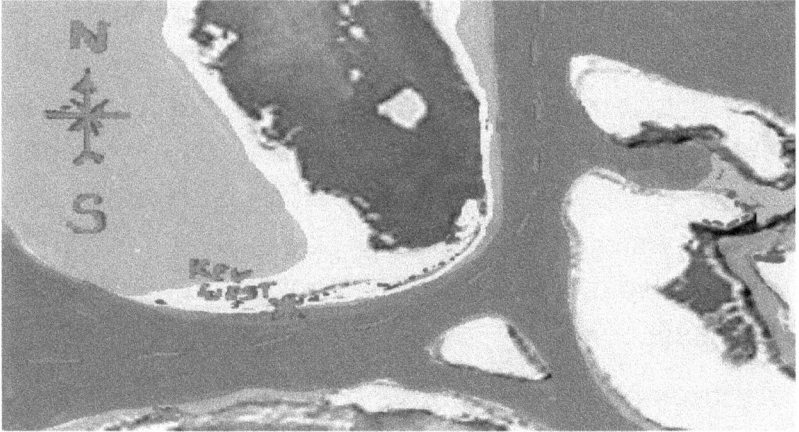

The Florida Straits

Over time more mainland areas were developed, commerce increased and shipping through the Straits became substantial. South Florida was a dangerous wilderness, near empty thus mariners scattered throughout the Bahamas were the first to develop as the early participants of the wrecking industry. As the various interests organized, the resolution of salvage rights between parties were generally settled in either Havana, or Nassau.

Wrecking soon became a widespread business and by the 1800s any Bahamian who could come by a suitable vessel at one time or another considered salvage and if successful, his salvaged goods were cleared and auctioned off in Nassau. New England fishermen often spent winters off the Bahamas and Keys selling fish to wholesalers in Havana. Recognizing a more profitable venture in which to employ their vessels, the Yankees tried their hand at salvage and before long, provided more competition than the Bahamian salvagers were willing to accept. Soon, war of sorts developed between the Yankee and English interests.

As it happened, at about the same period, in 1819, the United States acquired Florida from Spain and Congress passed a law a year later. Beginning in 1820 it was required that the value of any wreck salvaged in U.S. waters be accessed solely by a U.S. court. Thus, on May 7, 1822, Key West was confirmed as an official port of entry and became the primary center for the salvage operations in Florida.

With the prevailing winds being easterly, the majority of the wrecks were on the Florida side of the Straits, subsequently most of the Bahamian wreckers soon moved their operations to Key West. Any cargo not trans-shipped was auctioned or sold per contract, often with a great deal of legal confusion, for there were no U.S. statutes, only the "US Common Laws at Sea" and opinions held by the Bahamian and Spanish Admiralty

Courts differed widely causing ongoing conflicts concerning legalities, ethics and all aspects of wrecking.

With Florida now a part of the United States, the US Navy recognized the eradication of piracy in the Gulf Waters to be its responsibility. The West Indies Piracy squadron commanded by Commodore D. Porter was sent to eliminate pirates, who were plundering in what was now US Waters. American and foreign shipping were to be allowed to function without constant dread of attack. Based in Key West Harbor during the fleet operations, Commodore Porter felt compelled to advise the Secretary of the Navy of the chaos being caused by the absence of written legal statutes in salvage disputes.

Perry's letter was received in Washington but before congress bothered to get around to it, the Legislative Council of Florida Territory, wrote a well thought out wrecking act, listing exact procedures, legal requirements, and what would be the required officials and interested parties.

Also, *Section #14, laid out illegalities and penalties stating such things as: That the making or holding false lights, devices, or anything with the intent to mislead, bewilder or decoy the mariners of any vessel on the high seas, whereby*

*such vessel may be cast ashore, or get aground---
If convicted, they be deemed guilty of felony, and
shall suffer death.*

In 1828, the U.S. Government finally established
maritime and admiralty jurisdiction in Key West, in
the way of a Superior Court. The wrecking
statutes were strengthened even further in 1849.
A judge appointed in 1839, William Marvin wrote
the first comprehensive book on wrecking laws,
and made great improvements on what had been
an adversarial system biased against the ship
owners. Other courts at St. Augustine and
Pensacola were located far from the hazardous
Straits so, the great majority of the salvaged
property ended up warehoused at Key West.

Maritime courts aside, it was held that if a
settlement could be agreed between the parties,
the case did not need to go before a court. A court
was only required in cases of disagreement as for
instance, a case where the ship's owner did not
agree with the terms that the ship's master had
agreed with the salvager or if neither agreed, in
such case a court was necessary. Often, another
party, the underwriter, wished to represent his
interests, and if appealed, a case could drag out
for years. As the nation expanded, shipping and
commerce grew. Soon a hundred ships a day
navigated the Straits with on average, a grounding

each week. By 1830 Key West was per capita, the wealthiest city in the United States, all due to the wreckers. Almost all wreckers performed work other than wrecking. Many were fishermen, also spongers, turtlers, in fact any livelihood that positioned them and allowed them to respond quickly when a ship came ashore.

By Federal statute, a wrecking vessel was to be equipped with any gear necessary to salvage ships and cargo. Inventory like heavy anchors, cables, chain, fenders, block and tackle steam pumps. They carried long boats and barges to get in close where the water was too shallow for the large wrecking vessel to float. They used smaller scout craft to survey and board doing the advance work before the large wrecking vessels arrived on scene. The big vessel often crewed over fifty men who were up to any task which included diving, making emergency patches to refloat them or keep them afloat until they were brought to a dock or railway.

Watchtowers were erected at settlements. Wreckers offered finders fees, to anyone who reported a shipwreck to them, and small vessels would often patrol. Most wrecking vessels would, after spending the night in sheltered anchorages, sail out at dawn to discover if a ship had grounded. When a wreck was sighted a race would ensue. Captain of the first boat alongside a stranded ship, by custom became the Master Wrecker. As such, he was responsible for directing the operation and deciding how many wreckers he required to salvage the vessel and he was able to choose which vessels would assist. This process would often put ten or more boats alongside a wreck by mid-day with one man in charge.

The ship's master could, if he choose, ignore custom and pick whomever he wished to be the Master Wrecker or he could refuse all assistance but unless his problem was in the nature of a not so serious low tide grounding, the ship's master generally wanted assistance as soon as the first wrecker arrived on the scene. The Master Wrecker then was responsible for saving not only the cargo but passengers, crew and if possible, of course, the ship itself. If the ship's hull was not breached, only sufficient cargo would be transferred to the salvage vessels to float the ship free at the soonest possible high tide.

Salvaged cargo and gear were required to be taken to the nearest port of entry, Key West, where it would be appraised. The federal court would then decide how much the wreckers would be paid for their services. It did vary, with the judge taking into consideration the difficulty or efficiency of the operation but mostly averaged at 25% to 30% of the cargo's worth. Cargo was usually auctioned so the island provided a large number of warehouses and auction houses located alongside or near the docks. To support it all, there was a healthy shipbuilding industry, with marine railways, extensive repair yards, sailmakers, chandleries and metal works. The individuals who really got rich though were the ship owners and warehouse owners and auctioneers. Money flowed.

As the laws developed, a salvage or wrecking vessel and its master were required to be licensed; and the vessels owner's names had to be recorded. Syndicates existed when various wreckers worked together; short term and long term. The most common was when vessels agreed to work a particular wreck together, the contract between them becoming null and void at the completion of the job. The other type was a somewhat long term or permanent association, the partners, pooling resources and functioning as a single entity.

More often than not, wrecking crews spent days and weeks of hard work with little to show for it, always waiting for the big one. A lot has been alluded to, concerning dishonesty, from outright causing of wrecks to petty skullduggery, and pilfering of cargo. Overall, this reputation was unfair, for the Key West wreckers were definitely not crooks. The profession had its grey-areas, and many practiced precisely at legality's edge and a bit more; still, all trades have their tricks. For the most part though, Key West wreckers were honorable, hard-working businessmen.

Many wreckers operating during the peak decades of wrecking were famous. One of the better known was the high-handed Jacob Housman, a hard knuckle, ethically challenged young man, not in the least bothered by bending the law for profit. Young Housman, given command of a family schooner, (his father being in the New England coastal shipping trade), decided to sail without permission, for the West Indies. He arrived in the Keys around 1823 by way of running his father's schooner on a reef. Though it suffered serious damage, he managed to get the vessel clear and sail to Key west for repairs. In took no time at all for him to become enamored with the wrecking trade, and with his own father pressing for his arrest for theft of the schooner, he quickly decided that this was the very business in which he could succeed. During the following sixteen years he became known as enterprising, and a bold wrecker but also infamous.

Jacob Housman's name was synonymous with unscrupulous transactions, and theft from wrecks. So distrusted was he in Key West in the year 1830 that he found it necessary to transfer his operations up the coast to Indian Key. He began purchasing land on the tiny island, and rapidly became the owner of a majority, some of which he began leasing to those who needed space. Before long he'd built warehouses, he'd got a post office, and managed to have a new county, (Dade) created with his island as the county seat. His legal misadventures, too many to mention, came to an end in 1840. On August 7th, Seminole Indians attacked the island, burning the buildings and killing Housman along with other residents.

It is not too unreasonable to suggest that Housman, more than any other man, was responsible for the dis-repute in which the wrecking industry was held at that time.

Others, some even better known were of an honest disposition. One name stands out, Bradish "Hog" Johnson, nicknamed thus because for decades, he managed to get a hog's share of the work. It was dogged and precarious work, carried out by the wrecker's as they first engaged in an all-weather race to a wreck, and did so at any time of day or night. On arrival, there followed desperate effort, unrelieved, bone-grinding toil as men struggled to save the vessel, the cargo and the crew. The wrecking schooners were tough and solid, sailed hard by indomitable men, men like Bradish Johnson, for a weak or wavering man would never have risked wild surf roaring over a reef on the back of a hurricane. When the cry "*Wreck Ashore!*" rang out these men boarded their schooners and cast off regardless. Bradish was the epitome of what a wrecking master must be.

Bradish Johnson was son of Dean Johnson, patriarch of a wealthy Long Island family, who was known as a sportsman and was at one time Commodore of the New York Yacht Club. In one incident, Bradish's father, a splendid seaman, responded to the wreck of the bark, *Elizabeth*. The ship had been driven on a shoal off Fire Island and was breaking up. He manned the schooner *Twilight* with a crew of volunteers and sailed out in the teeth of a strong gale to attempt a rescue. With daring and consummate skill, he handled the schooner under impossible conditions, bringing her close aboard the *Elizabeth*, and saving many of her passenger's.

Bradish's mother, Helen, was related to Edward Lloyd, the first American yachtsman and she also had an affinity for the sea. With his father's example and a family heritage such as this, it was not a strange thing that Bradish Johnson was drawn to the sea and he entered the US Naval Academy in 1863.

Bradish was a good student and natural leader but balked at a life he considered to be regulated by foolish ceremonial embellishment and unquestioned naval discipline. He resigned a year before graduation and shipped out as mate of a New York to San Francisco, steam packet. After this job, he formed a partnership with his brother, in coastal shipping and commanded company sailing cargo ships on the Pacific Coast. Later he spent years in the seal trade and running guns to Mexican rebel, Porfirio Diaz.

Eventually, a friendship developed between Bradish Johnson and Diaz, and he was awarded the salvage of a gold laden ship, which had sunk in shallow water near a Mexican village. This expedition led to Johnson's near execution when he was captured by militants in opposition to Diaz.

Johnson returned to the east coast after the death of his brother, where he was awarded a contract by Admiral Perry to build a pier and light house on the island of Key West. The two things that Bradish found he liked best in Key West were the climate and Irene Bethel, daughter of the "Nassau Bethels" and one of the town's leading family. He bought a home, married Irene and in no time had earned a place among the leading citizens, those making up the town's inner circle. Leaving government contract work, he partnered up with the Baker Wrecking Company. Captain Ben Baker, a Master Wrecker, had been a master wrecker on a majority of wrecks, in fact, he'd been the leading wrecker of the previous two decades.

All Key West interests then focused on wrecking. Bradish Johnson's background, training and wide experience at sea brought him rapid acceptance as one of the industry's leaders. He became known for his audacity and relish in taking the long chance. No problem puzzled him, he owned a formidable physique, and was equal to any task he set himself to. Not only wrecking, for he'd built the 35-ton schooner *Irene* and another the *Island Home.* He took on jobs, contracted for projects, speculated on cargos, and set out on any adventure that turned an ethical profit, though he had no use for laws that seemed stupid or unjust.

When Baker Wrecking was bought out by the Merritt Chapman Company of New York, Johnson formed a consortium of his own, The Key West Wrecking Company. His partners were the Atchison brothers, Alfred and William as well as Peter Knight. These three men were Key West's best salvors and the fact that they joined up with Johnson shows the opinions held of his abilities.

During the following two decades their group took the lion's share of the work, salvaging up to four ships a month.

Names and numbers are all well but the work itself was often a desperate thing accomplished by bold men taking insane chances---and all under sail without the aid of modern navigational devices. One such story, concerning Hog Johnson, was told to me by an ancient salt who frequented my father's tavern. I was then, a boy of ten. My family had just moved south, and we lived off 17th avenue by the Miami River. I was then and have always been sea struck. I had paid rapt attention to the old man's story.

The Wreck of the Havana Packet.

A big steam axillary schooner, a Packet carrying general cargo and passengers on a regular run between Jacksonville and Havana had the misfortune to encounter an early season hurricane on her voyage south. The storm was tracking between Cay Sal Bank and Cuba and while in the apex of the Gulfstream, her steering had carried away. This happened in the late afternoon, and with a gale of wind from the SE, she was driven at good speed toward the Florida Reefs. Being in the vicinity of Santana shoal by dark, her crew did their best to direct her course parallel to the coast by the way of balanced storm sail to bring her bow up into the wind. This was successful for a time, until in a violent squall the mizzen storm tri-sail carried away, causing the bow to bear off to leeward toward the reefs. The crew were working madly to raise the peak of the mizzen gaff when the lookout reported breaker on the port bow. The master ordered the engine backed, hoping to slow her progress toward the reef, thus providing time to get away an anchor. Breakers reared up port and starboard as the anchor was cleared but it had no more plunged into the froth when the ship struck. She rose on the next great wave and surged, grinding across the reef until quite suddenly the anchor caught, bearing her around until she stopped, stranded, lying beam to the seas. The bos'n managed soundings and found the ship having sailed with a draft of sixteen foot, to be in less than two fathoms of water. In the pitch black, a pause in rain revealed faint lights visible to the north. The captain ordered two rockets fired immediately and then one to follow every ten minutes. A flare was to be kept burning at the mizzen truck. The wind was coming around, now from the south, south, east.

With dark and the murk of salt haze blurring the horizon, visibility was worse than poor, even from the top of the 90' tower that rose up from the SE corner of Key West. The lookout did see the flare go up. He got a compass bearing on the second flare, 161 degrees. That would put the wreck somewhere between West Sambo and Middle Sambo.

He shouted "Wreck Ashore" into the voice tube, as well as the compass bearing. The man stationed below mounted his horse and galloped toward town.

The Lookout Tower

Called out, the first of the wreckers to arrive at the bight, agreed that in the face of a hurricane there was not much hope for those on the wreck.

Stroud was heard to comment, "Unless the storm relents their likely doomed, poor souls. If she holds off breaking up, we might get a boat to them after first light -- if she lets up some."

Bradish Johnson hardly paused in passing the shelter of the Harbor Master's shack.

Appleby called out, "You're not meaning to go out sir?"

"I have it in mind," Johnson replied. "In this kinda blow," Appleby stammered, "even if yer able to haul to wind and get there, no boat can survive the breakers you'll find out there let alone do a job of work among them."

"Buck up George. I've worked wrecks in winds which this is by comparison but a mere zephyr," Johnson said with enthusiasm and offered a wicked grin as he turned back toward where his cutter waited.

Weather be dam'd, Bradish Johnson had his tough forty-foot cutter underway within thirty minutes of hearing of the stranding at the Sambos. His big schooner, the Irene would follow as soon as its crewmen were mustered. Other wreckers watched as he hove for the wreck and some few decided they weren't about to be cut out of a fee regardless of conditions. If Hog Johnson's boats were bound for sea, theirs wouldn't be found idle at the wharf.

Under triple reefed main and a club footed storm staysail, Johnson's cutter tore out of the bight, hardening up and passing Ft. Taylor close hauled. Clearing the island's southwest point, the cutter covered in blown spume, was vaulted up over one wave only to dive beneath the next, as her crew encountered the full force of storm. Guided by the rockets and later the flare, it required two full hours of tacking, to drive the cutter the mere five miles up to the wreck. Sailing to wind in over forty knots is generally not thought possible and is at no time pleasant. There was not a dry thread aboard as men pumped regularly to clear the bilges.

Coming up into the lee of the wreck on a port tack, Johnson had a man taking soundings. They found a depth of forty foot of water shoaling to thirty within a distance of two-hundred ft down wind of her bow. As the ship lay, she had a port list of twenty degrees but rolled further when lifted by the bigger seas. If her hull had been breached, and she was swept off the ten-foot mound on which she was presently stranded, she would sink in seven fathoms.

Johnson's Cutter reefed and rail down.

Her anchor was set leading aft to starboard and with the shifting wind, her stern might be moved. It would then be likely she would pivot on the anchor and slip into deep water. Not a good outcome for passengers or cargo. He'd needed to board the packet immediately. Getting the cutter close up under the ship's lee as it's canted rigging allowed, they set the cutter's anchor. Johnson then ordered the dory put over.

The stranded Havana Packet

Handling a leaping dory in a gale of wind is tricky, even when close alee of a wreck but it was second nature to these men. John West, another powerful man, went to the oars with Johnson in the stern. It was only a two hundred ft pull to windward, West rowed vigorously as Johnson baled and it took a few minutes, then more minutes spent timing the back surge to keep dory from being sucked against the wreck. A line was cast to the dory and at precisely 00:45 Johnson clambered aboard the Packet.

The Packet's captain waited at the rail to make introductions and to shake Johnson's hand. A wave then caused the whole ship to shudder and both men ducked behind the deckhouse as spray and green water swept past. Salvage terms were quickly agreed, and as Master Wrecker, Johnson set out with mate and engineer to assess the condition of the ship. As they passed aft to enter the machinery spaces, he observed through the saloon door, dozens of white-faced passengers huddled in bulky cork life vests. Arriving in the engine room, he was shown at the turn of the bilge, where wrens in the planking sprayed water each time a wave slammed into the ship. Due to its angle of heel, the damage was presently above the waterline and men were making what repairs were possible. The mate explained that the damage continued into the hold though, where due to the press of cargo it couldn't be reached. For the moment at least, the big steam pumps were staying ahead of flooding. Johnson returned topside to confer with the ship's captain.

He explained that the greatest peril lay in the ship being lifted by seas and tide and then driven into deep water. The ship being unmanageable must then sink or at best be driven aground a second time nearer the coast. He explained that the attempt would be made to place additional anchors to windward to thwart such an unfortunate outcome. Once anchors were placed to windward, an attempt to bring the lines down to the packet would be made, and her crew must be ready to bring them aboard and bend them to winches. Having agreed on what was to be attempted,

Johnson gripped a dangling line, stepped off the rail and swinging out, dropped into the dory. West cast off the long painter, and the dory was blown rapidly down to the Cutter, to find Tom Rooke's little schooner anchored within sixty foot to her starboard. At times she was pitching her bowsprit under.

Using a speaking horn, Johnson called to Rooke, "The captain is agreed, and I mean to set anchors to windward. I have a single 150 lb. salvage anchor below. What do you have aboard?"

"I have a 120 lb. navy and 100 hundred fathoms of rode," Rooke replied, shouting into his own horn. "And how do you mean to place them?"

"On port tack, sail to windward of the wreck and cast them over; fall off downwind across her stern trailing the lines which we cast off with buoys. If necessary, a boat to tow them close aboard."

"Will they be enough?" Rooke called.

"I think not" Johnson replied, "but the *Irene* is underway, and has aboard, four 300-pound anchors; two should be sufficient."

The task at hand, being understood between them, each set his crew to preparing the anchors and rode. Rooke's salvage anchor, being as a rule, lashed at his schooner's stern, left small work to rig, thus his was the first vessel underway. With a storm tri set in place of the schooner's main, they hoisted only a triple reefed foresail in order to break the anchor out. Expertly, Rooke angled away from Johnson's cutter as he sailed his anchor off the bottom; the vessel heeled severely, and coming around, jibbed. She scudded off before the wind, her crew, moving quick as cats, set and sheeted the storm staysail before the anchor could be winched up into the hawsepipe. Hardened up on a starboard tack, the little schooner leaped and plunged; cloaked in spray, she fought her way seaward of the breakers marking Middle Sambo.

Working mostly by feel in the pitch dark, Johnson's crew manhandled the cutter's salvage anchor out of the hold. Seas coming around the Packet caused a lively motion aboard and it was with some difficulty they shifted it aft and secured it outside the transom rail. The rode was coiled aft the helm and lashed in place, with the buoy tied off, also outside the rail, so to offer no chance of gear becoming fouled if boarded by seas. All in place, Johnson's cutter cleared away, following Rooke's schooner seaward.

The wind was now round to the south-south-west, continuing to blow at least forty. Though the seas were in the nature of 15' to 18' in deep water they were rearing up, breaking on the reefs to windward of the wreck. Staying in a deep channel, the cutter hove in the direction of Middle Sambo. Here, the seas had a regularity and she weathered them well. South and to seaward of the wreck though, there was only a mass of breakers where the bottom rose rapidly from a depth of ninety foot to a mere fifteen. Johnson would need to pick his spot to cast over the anchor, then cut downwind, streaming the anchor line as he ran the surf breaking over those shallows. It wasn't a practice any mariner would recommend. In this instance, Johnson's only other option would have been to cast over the rode with hopes the buoy would carry it clear to the ship and not fowl on the coral. Little chance of success with that. Rooke, also knew the score and would attempt to pull off the same maneuver.

Johnson watched as Rooke's schooner fell off and gathering speed was brought sharply into the wind and tacked about. She was swept, and staggered, laying well over as she came around but straightened as she regained her forward momentum. Now on a port tack she passed the cutter four-hundred foot to windward.

The cutter, being superior in its windward ability, was in position to come about no more than two minutes after Rooke's schooner had tacked around. Picking his moment, Johnson brought her about smoothly and followed, only angling a bit more to windward. He could just make out the schooner ahead on the starboard bow. He commented to West that Rooke was cutting it a bit close when a great-sea rose-up. It broke across the schooner's waist and carried her surfing sidewards more than a hundred-fifty foot to leeward, only halting at the very edge of the white water. She bore up and gained back fifty feet and then judging the distance right, Rooke turned and bore off downwind toward the stern of the wreck. The anchor was cut away as the schooner entered the breakers, its rode paying out, whipping off the stern as buried in foam, she made her run. A hundred-foot seaward of the wreck's stern, she overcame her rudder and broached to port, going on her beams ends for a moment before rounding-up with her nose to seaward. Sails in disarray the schooner swept stern first past the packet's counter, but the buoy floated inboard and alongside the packet's hull. The anchor line was in reach of the Packet's crew.

There were a few among the twelve men aboard the cutter, men who'd watching Rooke's run, had cringed and these souls were unsettled in the knowledge that their trial was at hand. There was nothing for it; wrecking was their trade. Johnson ordered the mainsail doused, so as to have no wind pressure aft as he began to angle downwind. A big sea reared up becoming vertical. He, judging the distance to the wreck at inside of 600 foot, yelled, "cut away." Even as the anchor plunged, an avalanche of foam and green water roared over the transom, swamping them to the waist and boosted the cutter forward at a mad rate. They were carried three hundred feet before the body of the wave passed, and the cutter was slowed by the back surge. Then, while only a hundred foot short of the Packet's stern, they were overtaken by a greater breaker. This sea lifted the starboard quarter high, levering the cutter's bow in toward the ship as she surged forward attempting a broach to starboard. Dick Lowe cast off the staysail sheet as Johnson held the tiller hard against its starboard stops. It was a near thing.

To the layman It may have appeared a neater run than Rooke's perhaps, but it was surly a dam'd closer call, for the cutter rushed wildly past the packet's stern, clearing with less than a yard to spare, and rounding up in the packet's lee. West kicked the brake and the cutter's anchor dropped away, the wind caught her nose driving her rapidly astern, until the hook caught, and yanked her bow back around into the wind.

Men took deep breaths and gave each other nervous looks, then began to furl and clear away. The dory was unlashed and put over the side before Rooke's schooner got back up and anchored. Watching her come in, Johnson passed round a bottle of whiskey. He noted, the packet's mate already had strain put on one of the anchor cables; the crew were currently fishing the other out from under the counter. Curses drifted down to him for the breaking waves were making this no easy task.

In the distance blue flares announced *Irene's* imminent arrival and two other small wrecking vessels, one of them Peter Knight's, were close at hand. Guessing *Irene* to be at least, thirty minutes out, Johnson spoke to Rooke, and determined that eight men from each of their boats should board the wreck. Working with her crew they would attempt to move her port anchor aft, dropping it at the stern, ready to be hove tight if the ship should begin to shift. Then they would attempt to move enough cargo to get access to the starboard bilges of the hold. A light warp was then stretched to the ship, so the schooner's dory's might be hauled easily to and from the anchored wrecking vessels.

It was now near three am, three hours to high water and a great deal yet to accomplish. The packet was a well-built ship, teak over iron frames but the seas bursting against and over her starboard side were giving her a hammering that nothing built of wood could long withstand. The *Irene* was now drawing close, and leaving Lowe in charge of the cutter, Johnson and West climbed into the dory, intending to drift down and board the approaching *Irene*. As they passed, *Irene* came up into the wind, slowing in an instant and the dory slipped under the transom. Catching a thrown line, they were pulled close aboard as the helm was put up as she began again to make way. Johnson scrambled aboard taking command as the dory was hauled aboard.

Directing the *Irene's* course to seaward of Middle Sambo reef, he motioned for his crew to gather round. He explained what their course of action would be, and what must be done in preparation. The wreckers turned to the work straight away, first rigging the big admiralty anchors, each with an iron buoy to float its rode. After the anchors and their buoys were dropped, the 22 ft whale boat would be used to tow a light sea painter to the ship. This painter would be used to heave the anchor cable to the ship. In no event did Johnson wish to run the surf a second time with the larger vessels.

Abeam Middle Sambo, the *Irene* tacked and was brought to a position a half mile to seaward of the packet, before being put about a second time and hove to with her staysail and jib aback. Drifting down on the wreck, both admiralty anchors, with their buoys and 10 fathoms of chain were slipped over at a depth of 4 fathom. Slow flares had been attached marking them. The sheets of *Irene's* head sails were immediately cast off and hauled to port so she might gather her headway and avoid sliding down into the breakers.

The wind had come near southwest by now and *Irene* had again been worked to weather of the anchors. To those men who manned the whale boat, would come a bonus and most aboard volunteered. Johnson chose 4 for the oars and one to bail, all but one unmarried, and would himself handle the sweep oar. The mate James Burton was to remain in charge of the Irene, with instructions to maintain a position directly to windward and pour oil on the water at intervals to reduce some of the winds grip on the surface. With all understood, the *Irene* was brought up near irons to afford a lee, and the whale boat was released on a roll. Without hesitation, men dropped into her fending off with oars and casting off her painter. Three men managed to get in a good pull of their oars and braced in the stern with the sweep oar, Johnson managed to bring the boat around before they came out of the *Irene's* lee. Oil being poured out of the schooner's wake was having a beneficial effect and blasted by a gust, they begin pulling for the buoys in good order.

The wind pushing them, being a strong gale, it required little effort to propel the boat, only the occasional hard strokes to pull ahead of a crest before it broke or to back water for the same reason. They soared up on the waves affording a good view of the flares which marked the buoys, then sank into obscure canyons. A band of squalls had passed allowing some light from a three-quarter moon. They were, so far, assured in their minds, confident in themselves and this boat, for other than being broadside or boarded by a breaker, there was little to threaten a well-handled whale boat. This being a wrecker's boat, with surf reckoned on, it had in its construction, sealed compartments in the ends and cork fixed under the thwarts. Even swamped she would float and though sluggishly would handle. In modern times, there are few seamen who can claim to skill or proficiency in handling a vessel propelled by oars in flat water, let alone in a high-seas gale. It was a rare skill even in the time of sail but this was still in the days of sail. Bradish Johnson and men like him were masters when it came to handling a boat of this sort.

Hog Johnson's Surf Boat Running the reef with the anchor Pennant.

They rose up out of a trough and it seemed as they were suddenly on top of the most westerly buoy. Steering close aboard, Johnson called for the men to back water while the fifth man attempted to secure the anchor pennants with a boathook. In the midst of this, a cross sea rose up and put a substantial amount of water aboard.

Never-the-less, the pennant's eye was dropped over a stout post.

They fell off with three men bailing, determined to empty the bilges in the minute before they entered the breakers. Man-o-wars, swept aboard with the seawater, burned and branded skin and over the whistling wind and hissing water, curses passed back and forth between West, Birdy and the other bailers. The oil was doing some good, yet now the surfaces were slippery, and It was reason for more complaint.

They were close to the break and Johnson called out, "oars, boys".

The first breaker was not so bad, but the comer that came on its heels was a ten-foot avalanche of froth on the face of a twenty ft wave.

"Back water," he yelled, then "hold water," as the stern shot-sky-ward.

They took it straight on the stern, Johnson's lower body like a bulwark parting green water. The drag of the pennant's towing aft helped as the wave passed beneath, some of it spilling aboard to deposit six inches of water in the bilge. Another comer was roaring down on them and there was enough moon to show a mix of emotions among the boat's crew at that moment: fear, exhilaration, and a mug or two that suddenly wished they were elsewhere. "Pull dam'd you," Bradish Johnson yelled. He bent, braced positioning the sweep and again aligned the Whaleboat with the tumbling wall, "pull! or be ready to swim," he shouted above the water's roar.

They counted ten powerful breakers in crossing the reef. Swamped to the thwarts and a mere fifty feet from success they reached the end of pennant line. The boat stopped dead and the next wave swept them, two men going overboard were dragged back aboard, one clutching his oar as he got a leg over the rail and rolled inward. They bent another line to the anchor pennant and cast off the eye.

"Pull and earn your shares," Johnson yelled and neck deep in the hissing foam of the next breaker they did.

One minute more and they crossed into the packet lee, heaving lines flying out to them; one taking Birdy full in the nose put him off his bench. They got hold of two and were hauled up just forward of the Packet's mizzen shrouds. Men climbed from the swamped boat passing up the anchor pennant line as they came. While some attended to emptying the whaleboat, Johnson saw to leading the all-important anchor pennant to a winch on the

Packet's Starboard quarter. By 05:00 the cables of both of the big anchors were put to strain. The wind had shifted to west-southwest lessening the hammering which the port side had been enduring and it had been necessary to move the smaller wrecking vessels toward the bow. Peter Knight and his men had boarded earlier and joined Tom Rooke in the Packet's #2 hold. They reported having removed a number of ceilings and wedging planks over the worse of the damaged planks. When weather abated enough to lighten the cargo she could be floated and towed.

Rooke commented, poking his head out the deck house companionway, "I do believe the winds off some."

"Seas are worst though, nasty, crossed with the wind shift," Johnson said. "I think we've seen the worst of it though. Be able begin littering by this evening, by the morning at the latest. Won't take too much to float her."

The stern of the packet lifted and sat down on her keel again, as she had begun doing several times a minute. Knight gave Rooke a concerned look. "Dam'd good thing we were able to get those anchors placed."

"No doubt," Rooke replied. "looks like this one will bring a dam'd decent fee. Lot of silk down there just to mention one of many goodies."

Knight smiled, "No doubt."

"We'll get him towed into Key West, but the Captain wants towed to Havana for repairs," Johnson informed. "Once the salvage is settled," of course.

"Passengers?" Knight enquired.

"The captain can deal with that, I suppose. He offered the charthouse settee so," he sighed, "I intend to rest my head." Johnson yawned. "Chances we'll discover a few more stranding's with dawn. Promises to be a busy week."

By 1890s more lighthouses were built and with stronger lights. The available charts were more accurate; a larger proportion of shipping was under steam power and less susceptible to being driven ashore in severe weather. The number of wrecks on the Florida Reef continued to fall and by the turn of the century the report of a wreck ashore was a rarity. On most of the world's oceans, sea-going tugs had replaced wrecking schooners.

On Ajax Reef off Miami, the stranding of the steamer *Alicia* of Bilbao is usually considered the final gasp of the old Florida wrecking industry. That was in 1905 and wouldn't you know, Brandish Hog Johnson was Master Wrecker. He was assisted by over 70 other wreckers swarming from far and near for a share of the rich cargo. After the better part of the *Alicia's* cargo was taken out, she was floated off the reef but hit by a violent squall, she sprung a leak and a short time later, down she went.

That was the end of wrecking and Bradish Hog Johnson, himself, passed on five years after. In 1905, while hauling out a schooner on his railway, he died on his feet, vigorous and hardworking to the end.

The Pilot Boats

The competitive nature of pilot boats; Second Goes Hungry.

The word "pilot" comes from the Dutch, in fact the tiny country of Holland which figured so largely in European maritime exploration and expansion in the 17th century, produced the first professional

Western European bar pilot of record. That Hollander was the enterprising Franz Naerebout who in 1749, let it be known that he was capable of guiding sailing vessels in and out of Dutch harbors. He advertised that by virtue of his personal knowledge of channels, rocks and shoals he could safely guide a ship to port. Until that point most old-world navigators had crept into and out of unfamiliar harbors, with only the aid of crude charts and a sounding line. The option of hiring a specialized professional was tempting. Being guided by a knowledgeable mariner familiar with his own harbors and its weather conditions was far safer. It was a service that captains in ports all over the world have come to appreciate.

Before the Dutch entrepreneur Franz Naerebout, the title of pilot referred to a responsibility. Pilots would guide ships or fleets over vast distances, staying with a ship for its entire voyage, even around the globe. This practice was ancient. We hear of it in the literature of antiquity. Homer's Iliad praises the pilot Faster for leading the Achaean ship safely through the rocks and shoals. And if a pilot failed--- Often the penalties could be stringent. For instance, in the ancient city of Rhodes, navigation law required at the demand of Merchant or master, for a pilot to pay for the loss of a ship or cargo or lose his head. Thus, pilots tended to be skilled. The writings of Marco Polo, credit was given to the skill of Arab pilots during the explorer's travels in the Indian Ocean. Vasco da Gama credited Portuguese and Arab pilots with much of his success.

During the exploration of the New World, the role of pilot was increasingly elevated and rewarded, frequently with title or government office. America Vespucci, the cartographer for whom two continents are named, was given in 1508 the title of *First Pilot of Ships of Spain*. This called for the preparation of all charts, using information gleaned from previous voyages and the report of travelers returning from far parts of the world. Another example was Sebastian Cabot, who on return from explorations of North America, was named and given the office of *Grand Pilot of England*.

Eventually, the term pilot, came to be reserved mainly for those with **local knowledge** of specific harbors, rivers or estuaries but even here the skill required an understanding of the handling of ships of all rigs. Getting to the and aboard the ships required other skill sets and specialized boats. Those developed in the treacherous waters of the English, and Bristol Channels were some of the best, the most famous were called Bristol Pilot Cutters.

Britain's Bristol Pilot Cutters

Originally based on fishing boats that had a simple one mast rig, pilot cutters evolved to deep hulled boats with a plum bow and elliptical stern. They were gaff rigged with the mast stepped 2/5 aft from the bow and had a long bowsprit, allowing for adding a large jib for speed. To dock or maneuver the bowsprit could slide aft onto the deck. Many early cutters launched off the beach but in time they became larger in order to carry a number of pilots out to the western approaches and remain at sea for days on end.

Bristol Pilot Cutters

To skipper these fast and nimble cutters, pilot associations hired only the best and most experienced sailors; each was required to have in-depth knowledge of local waters to be able to put a pilot aboard a ship in all conditions. On the smaller cutters, often there was only a pilot and apprentice. Once a pilot was aboard, the apprentice takes charge of the cutter and single handing, follows the ship into harbor to retrieve his master.

Always seeking to beat out a competing pilot cutter, (*there were a lot of pilots*) was of major importance. With limited employment opportunity, it was survival of the fastest but speed and maneuverability weren't the only consideration. The English Channel and Bristol channel deserve their reputation as being two of the most hazardous of the world's shipping areas. Fog, current, nasty weather, confused sea, and extreme tide ranges. The Bristol channel has perhaps the planets greatest tidal range. Its over fifty ft with copious rocks and miles of shifting sand bars. The conditions of that environment require an extraordinary boat. sea, and extreme tide ranges. The Bristol channel has perhaps the planets greatest tidal range. Its over fifty ft with copious rocks and miles of shifting sand bars. The conditions of that environment require an extraordinary boat.

Cutters on the flats at low tide.

North American Pilots

On the Northeast coast, the approaches to New York City had an ardent competitive spirit and the men engaged in piloting were capable. One pilot who can be mentioned as an example of that sailing skill was the Sandy Hook Pilot, Captain Dick Brown, who gained renown when he outsailed the yachts of Great Britain to win the first, *America's Cup*. To earn their fee, Brown and his old school colleagues must enter and depart harbors under sail in ships of varied size and efficiency, and do this through many channels, while dealing with winds and tides that could range from uncertain to perilous.

Before the year of 1837, there was no organized system of pilotage for the Port of New York. The winter of 1835-36 was a disastrous one for shipping on the northeast coast, though. One tragic incident drove the officials of the port to take action.

Late one winter day, a packed immigrant ship hove too off Sandy Hook Lightship with signals flying for a pilot. A gale was building from the southeast and with no sign of assistance, the crew began to fire cannon to attract one of the pilot boats at anchor inside Sandy Hook. Still no pilot had put out and by dark, the ship was in a desperate situation. Unable to find the Narrows channel or to beat to windward against the gale, the ship was driven onto the Long Island shore. By dawn, over 400 men, women and children had been drowned or frozen aboard the wave swept wreck.

This catastrophe brought about the appointment of a Pilot's Commission for both the States of New York and New Jersey. With two separate Pilot's Associations, the competition became lively, and even keener when the New York merchant shipowners, and insurers demanded that a third group of pilots be commissioned and that all cruise well out from Sandy Hook to meet incoming ships.

Next, the State of New York added the real competitive factor. They legislated a further fee as an incentive. It amounted to an additional 25% of the total if a ship was picked up beyond 15 miles of the light. That encouraged the pilot schooners to cruise as far out as Sable Island, well over 600 miles, to meet ships. So intense was the rivalry between pilots operating out of New York and New Jersey ports in the late 1800s that the pilots took their boats as far out as the Grand Banks vying for the piloting fees.

t was often a breakneck race between pilot boats to sight incoming ships and be the first to hail. Many thousands of hotly contested races ensued, each starting with the first glimpse of sail on the horizon. A combination of sport and business, perhaps a little madness at times, but they pressed their boats day and night in all seasons and weather. Many were lost and victory was only to allow the pilot to do his job and earn his fee. The pilots on the losing boat got nothing, less than nothing, for they were often far to sea and hours behind in the next contest for the pilot's fee.

A bureaucrat, Palmer Campbell spent much of his career as director of large East Coast ports. In a conversation, he recalled vividly his own experience with the adventure and perils of the piloting industry. As a boy of 12, his family had booked passage on the SS *Calabria*, a packet of the Cunard line. From the liner's deck, he had witnessed a spirited duel between two front runners among the pilot boats racing toward the SS *Calabria*. They had carried full sail and were being driven hard as they vied to reach the ship first. It was an experience he would never forget.

A half dozen liners and packets had been scheduled to arrive, and boats from several pilot associations had moved further offshore during the night, all positioning themselves to get the jump on competitors. Since dawn, lookouts aboard pilot vessels had scanned the horizon for smoke or sail, for any sign of an approaching ship. A stiff breeze was blowing, and double reefed, on starboard tack, the schooner *Phantom* was about four miles to windward of three competing pilot schooners; each jockeyed for what her captain deemed a favorable location. The nearest, a new spoonbill schooner came about on an intersecting tack shortly after 08:00. An hour later the *Phantom's* look-out spied a whiff of smoke to the north. Her captain brought her around on a broad reach immediately, calling for the crew to shake out a reef. Within a minute the spoonbill, finding itself in an equally favorable position and in fact, a little closer, followed suit and also fell off a heading to intercept the ship. The race was on.

Swiftly, the schooners angled across the face of the swells surging forward with foam sweeping the decks, quartering seas lifted their counters. Within ten minutes, the *Phantom's* master, realized she was overtaking the spoonbill, but he feared his speed was not fast enough to overhaul her in time. He ordered the last reef shaken out of the mainsail and both topsails set. He drove his vessel ruthlessly, rigging humming, decks awash. She flew, swept with foam, she ran like a whipped horse toward the horizon and *SS Calabria*, whose smoke was now plainly visible from the deck.

Thirty minutes and the spoonbill had hoisted a large fisherman forward of her main, boosting her speed. Because of this, she still held the lead. The pilot schooners were neck and neck, but *Phantom* was continuing to inch-up. They were confident they would pass but even so, the ship might already have picked up a pilot further to the east making the race a vain effort. Some schooners ranged as far out as Sable Island, some 600 miles from New York. A mile off the ship hove too and vented, a sure sign that a pilot was still required. Within half that distance *Phantom* forged ahead, and as she closed the ship, plainly crossed the spoonbill's bow. A pilot aboard the *Phantom* positioned forward, raised his arm; On the *Calabria's* bridge, the captain raised his arm in reply and pointed in *Phantom's* direction. The race was won; her pilot would be taken aboard.

Double Reefed Pilot Schooner on Station

Astern, the spoonbill boor up in a graceful semicircle, beginning to reduce sail as *Phantom* also reduced, at the same time prepared to launch her dory. After watching the two schooners, tearing toward them from over the horizon, sports minded passengers on the deck of *Calabria* were giving up rousing cheers. Good sport or not, for the spoonbill, it was a case of Second Goes Hungry

.To persist in the race, and earn a profit, a pilot boats had to be built not only for speed but for endurance. Stamina was a prerequisite in the brutal North Atlantic waters. Few of the sailing era

New York pilot schooners or their pilots lived to retire; most were lost to the sea. In addition to the perils of their own profession they challenged death in attempting historic rescues. In March 1886, the pilot boat *PHANTOM*, mentioned earlier in this section, rescued all 700 passengers and crew from the *S. S. OREGON*, which sank off Fire Island within sight of land. Aboard the Phantom, all space below and every inch of deck room was jammed with survivors. The valiant little vessel received the fullest recognition from both shipping interests and the government. The *PHANTOM* herself and six of her brave crew were lost two years later in the blizzard of 1888.

Pilot rowed to hove too clipper

Grand Banks Fishing Schooners

Two Grand Banks schooners in the 1939 Last Race.

Downwind under storm sail.

For several days' captains had scrutinized the waterlines of nearby vessels, mainly those they judged to be near full and down. A few captains, those with near to a full load, made sure their own schooner would be ready to hoist sail and bolt for the coast in a moment's notice. East coast fish houses paid a premium price for the first fish of the season. If your hold was full or near full it was a fine bonus.

Two man 12' Dory crew

Mid-morning, and a fog had begun to set in with a freshening breeze. The *Nannie Bohlin* had been observed, setting his dories late and had recalled them at the first whiff of fog. Captain Bohlin was a prudent man, yet---his actions seemed a bit over cautious for a day when the fishing was good.

Ben Perry, who was pulling hard for the sound of the *Elsie's* horn, was the first to figure they'd been foxed. Twenty yards to port, the silent *Nannie Bohlin* slid past. In seconds she was only a shadow fading into the mist.

"*Why, devil take the sneaky skunk,*" Perry cursed.

His dory-mate, Caleb Smit, nearly as astonished, cut loose his fish and plopped his buttocks onto the after bench. Dropping his oars between the thule pins, he timed it and leaned into them to add to their speed. "Captain be some steamed," he muttered.

Four boats were away yet when Perry and Smith hailed *Elise*. The Captain, previously, only concerned with fog and a rising wind, now did all but turn inside out in impatience to get his dories aboard and be underway. It was near an hour before the last dory was hauled aboard. The *Elise* cut out in the *Nannie's* wake and did so every bit as quietly.

Sailing Rail Down

The wind, still rising, had all the appearance of building to a gale. *Elise* reached through the cold soup of fog and driven rain under full sail, flying all but the captain's long johns. Rail under, they drove through the night, helmsman knee deep in the white water, grasping the kicking spokes, eyeing the bend of the main boom, wondering if it would break and gambling it wouldn't. By dawn the visibility had cleared some but for that, the wind blew the harder. *Elsie* was to wind and a bit astern of the *Nannie C Bolin*. The gale was around to north northeast now, and they ran off one long sea to dive under the next. Broad reaching down the grey rollers, the Elsie's speed surged and whole waves at times rushing over the deck, the rush of water competing with the howl of wind through the shrouds.

Laid over to the second ratline, with water up to the hatches, *Elise* heeled before the power of the storm. *"Shorten to the first reef!"* advised old Angus McHanna, the *Elise's* carpenter. He shouted over the wind, *"Dinna be daft! She'll go the faster with a tuck or two in her!"*

Laid over to the second ratline, with water up to the hatches, *Elise* heeled before the power of the storm. *"Shorten to the first reef!"* advised old

Angus McHanna, the *Elise's* carpenter. He shouted over the wind, *"Dinna be daft! She'll go the faster with a tuck or two in her!"*

Heeled to the hatches

The captain gave the nod, but it was no easy task to accomplish. By the time they'd fought abeam the Nannie C Bolin, wind had topped 45 knots. Great waves foamed off to leeward, fading into scud so thick it obscured her hull and it was about then, the *Elsie's* staysail exploded. It burst with the report of a canon shot and for an hour crew struggled to set new canvas. Touchy, for the occasional wave over-topped the bow and took men off their feet. They got-her done through but they'd lost a bit ground. Main triple reefed and standing like an oak, the *Elise* finally lurched past the *Nannie* some-time around noon. The Elsie's crew howled a shout that could be heard a mile to lee. As *Elsie* continued to pull away, her captain chuckled and calculated the extra cash price of the first cargo on the wharf. Not that much higher, perhaps, but Oh, the glory! Aye, but then--- they weren't there yet.

Neck and Neck

Going back a thousand years

The Grand Banks had already been a prime fishing area for Basque fishermen, hundreds of years before the first Gloucester men sailed there in 1657. Norsemen, led by their intrepid Erik the Red and his son, Leif the Lucky, fished there in the 12th century in open longboats with handlines made from horsehair and sheep's-gut and hooks

Scandinavian Knarr fishing vessel

fashioned from bones. Norman, Breton, and Basque fishermen almost assuredly began fishing Newfoundland's Grand Banks as early as 1497, the year John Cabot explored the area and commented on vast schools of fish. These Grand Banks are shallow places in the North Atlantic where plant and animal plankton thrive and feed huge schools of codfish. The Grand Banks were among the best fishing grounds then known, and Frenchmen were among the first to exploit them. The first to fish there kept the location secret but reliable records show that fishermen from the channel port of Honfleur fished the Grand Banks as early as 1504. Other records show that boats from Dieppe were there by 1507.

Basque fishing vessel around 1450

1731 there were 160 fishing vessels sailing out of Marblehead, Massachusetts alone. Gloucester, too, was a vital fishing center and by 1788, at least 60 vessels from Gloucester were harvesting the Grand Banks. Fortunes were made in the fishing fleets. One Salem entrepreneur, Benjamin Pickman, built himself an imposing home in 1750 and commissioned artisans to carve codfish on the end of every stair in the front hall in frank and grateful acknowledgment of the basis of his good fortune.

During the Revolutionary War, the New England schooners abandoned fishing off the Grand Banks because the British men of war patrolled the area, sinking the fishing schooners and pressing the American crews into British service. In the treaty ending the war, Americans were given rights to the Newfoundland Banks but were not allowed to dry or salt their catch on Newfoundland soil. This meant that instead of going out for a stay of six months to catch and dry and cure their fish in one outing, the sailing vessels had to take their load of wet-salted fish back to New England ports and dry it there. After much argument on this score, John Adams oratory finally won the day when he addressed the treaty delegation in these terms: *"When God Almighty made the Banks of Newfoundland, at 300 leagues distant from the people of America and 600 leagues distant from those of England, did he not give us as good a right as he gave England?"*

Some schooners brought their catches home live in a "well" of seawater amidships, others cleaned and layered them in ice, or salted and dried to an almost indefinite state of preservation. The use of ice to preserve the catches was begun early in the 1800's. Sailing vessels brought the ice from Norway and the United States, particularly from Wenham Lake, just west of Gloucester. Each schooner captain selected his fishing site on the Banks by experience, educated guess. First, sounded for depth using a lead filled with tallow to discover the nature of the sea bottom. Then, considered the weather, the recent gales, and the current. He imagined the environment as a fish searching for food would. If the tallow collected mud, the experienced captain knew their would-be poor-quality fish like hake in the vicinity. Fine sand would mean haddock, which could not be salted, but if the sample was coarse sand with bits of shell, he knew it was the spot for cod and the schooner settled down for weeks of fishing.

Under the threat of death and disaster as the fishermen constantly worked, they developed a wealth of superstitions: no women on board, no use of the words "rabbit" or "chicken", no mention of salmon (although other fish were the main topic of conversation), and no talk of corpses. It was particularly bad luck for the crew to find out how many fish they had caught. Only the captain and first mate could possess this information with immunity.

Early in the 19th century, the Grand Banks fishermen began leaving the schooner in 12-foot dories to work their handlines over a much larger area.

Grand Banks Schooner near fog banks.

Every morning a fleet of dories would set out from the mother ship, returning in the evening with their day's catch. In good weather, the dories were tied up to float during the night. If the seas were rough, they were hoisted aboard and stacked overnight.

The dory men were the backbone of the vessel and they looked on anyone who fished from the deck with a certain professional disdain. In pairs or at times alone, they set out in their open dories daily hazarding the open ocean waters of the Grand Banks. Only heavy fog or strong gales could keep them aboard the schooner, where they then swapped yarns and mended their gear. Always, men's eyes were on the weather. If dawn brought decent weather, they swarmed away from the schooner, every man taking his dory in the direction he chose. Each man carried his lines baited and coiled in a tub, a little ship's biscuit for nourishment, a jug of water and a conch shell to signal the mother ship. Some of them rigged small sails on their dory, others rowed. Away from the schooner, they were quite on their own. If a man fished hand lines, he knew his catch, but a trawl had to wait. After fishing all day, the dory man stood up wearily, often wet and cold, in his rocking little dory and pulled up the anchor and groundline to learn his fortune. Many hooks would have been emptied by predators, some would hold worthless fish or small sharks. Sometimes he would find a valuable halibut and usually many codfish to help fill up the hold of the schooner, maybe allowing it to cut out early for the long haul back to port.

On occasion, a rogue wave would swamp a dory and dump half a ton of cod as well as the dory man. Bundled in his thick woolen clothing, heavy boots and oilskins, the man could not remain afloat and, if he could, would not survive in the icy water. He had to scramble back and bail her out; keep moving so the friction of the wool would bring back his heat. Quite apart from these perils, there was the ever-threatening summer fog, which would appear like a low, grey cloud on the horizon and approach at incredible speed, enveloping the dory in a thick shroud. Then the dory-man had to pull in his gear fast and head for the mother ship, blowing on his conch shell for signals. With luck he would hear the answering blast from the schooner. If he did not find his way back, or get picked up by another boat, he lost his life to the sea or perhaps landed on a coast scores of miles away. More than one-man rowed hundreds of miles to the coast to save himself, often having all his fingers frozen off in the process.

Through all the perils, the sailing Grand Banks fishing vessels and their indomitable dory men survived into the 20th century. The last American dory schooner was the *Marjorie Parker,* which was put ashore on the Fairhaven Bridge near New Bedford in August 1954 by Hurricane Carol. Now it's all diesel and trawls.

The Smugglers

Profit or Prison

Night, low racing clouds, dim, the dark side of the moon barely visible. Ahead is a narrow pass to the open sea. A fore topsail schooner, all canvas set drives for the pass and the safety of the open sea, while a revenue cutter, charges spouting smoke and fire to cut her off. Fortunes are at stake, action, adventure, romance, the stuff of books and movies but it was once real.

Ah, it's not always the gain: there is the excitement of the getting. This way of life is as old as avarice, a sort of deviltry that has always stained the fabric of proper society. Strange, how many of the pillars of society have made their mark first on the wild side. The smugglers, pirates and scoundrels have woven many a spicy tale into society's colorful tapestry. What boy hasn't dabbled with visions of sailing in with a secret cargo; what man hasn't considered profits won from contraband passed ashore in the dark of the moon or the thrill of outrunning the revenuers astern?

Who wouldn't rather plunge into a tale of treasure, jewels, captive beauties and daring do than a staid gothic drama? Courage and audacity stir the heart, regardless if their possessor is hero or villain. It is the spirit of such souls that recommend them. Be it glory, greed or kinship that draws them to some hidden strip of sand beach, requires torturous escapes through narrow rock-strewn channels, we admire their willingness to engage in daring battles against all odds, refusing to be discouraged and challenging fate itself to attain their goals.

Smuggling most likely (who could know) dates back to the first time a sovereign did or did not want something to enter or exit his territory. One example is the Emperor of China who did not want silkworms to travel anywhere; opposed was the Byzantine Emperor, who was interested in having silkworms travel to his lands. In the end a couple of very well-paid monks managed the necessary smuggling and Constantinople soon monopolized the production of silk in Europe.

Smuggling is often related to pains taken by authorities to thwart the importation of contraband goods, both taxed items and product forbidden for various reasons. During the reign of Edward I of England, smuggling first became a documented problem in 1275 CE but only after he initiated a customs collection system. The smuggling of highly taxed export, and import goods became immediately evident. Endless things were smuggled for limitless reasons. The Customs and Revenue service on one side, the smugglers and merchants on the other. The only records detail the activities of those unwise enough to get caught in the act. The others got rich.

For such activities, be it for hunter or hunted, specialized vessels were needed. On lakes, rivers or wide channels; skiffs, gigs, lugers, or galleys. On seas, gulfs or bays fast and handy sailing ships were the ticket; nimble vessels that could turn in their length were the very ones required by scoundrels as well as the men who were tasked to hunt them. From such need was the Baltimore Clippers, and fore topsail schooners brought to perfection. Yes, during war they made excellent privateers but smuggling went on year in, and year out during war and peace. Be it spirits, arms or slaves these were the perfect choice for smuggling. One wealthy New York ship owner, and member of the yacht club, sent his fast schooner to England with cotton, from there to Africa with arms, and to Cuba with slaves; all the while avoiding or outrunning British anti slave patrols. Back to New York, with a cargo of fruit, the vessel was given a though cleaning, then raced off long Island during the season. This guy had to have been one efficiently ruthless businessman.

For centuries, crude forms of the schooner rig evolved in far corners of the world but in the 16th century Americas, small schooners became generally the most popular rig due to ease of handling and windward ability. The American revolution and the war of 1812 that followed spurred the development and increase in size when they became favorites of both privateers and coastal navy patrol vessels. Builders in the Chesapeake Bay ports developed what became some of the fastest vessels of their size in the world, which came to be known as Baltimore Clippers.

They served their owners on all sides of conflicts, smuggling, running blockades, naval patrol and dispatch. As privateers, where the speed and maneuverability were of the utmost importance, they were armed with light cannon, carronades and swivel guns, perhaps a heavy long-gun midship. Sailing with a large crew, they could swoop down on a convoy, cut out a few merchant ships, and escape to windward before the more ponderous naval escort could do anything about it. Some owners had them built as slavers, or to carry other perishable cargo.

In the decades leading up to the American Civil War, the Baltimore Clippers as the class was called, were some of the fastest ships in the world and their hull shapes began to influence the design of the largest square-rigged merchant ships being built at that time. First the Extreme Clippers were built to carry miners round the Horn to the California gold fields, then the smaller Tea

Clippers followed. The fisheries of New England also honed both hull and rig to their own requirement.

The swift smaller schooners, even past the peak of their dominance, made good use of moonless nights and hidden channels. And with the advent of prohibition, led the first forays against the laws, customs officers were tasked to enforce. The federal government gave Customs and the Coast Guard their orders but with a lengthy coast: 12,000 miles on the Atlantic, Pacific and Gulf, plus the great lakes, it was an impossible job. When the Volstead Act came into effect, there were only 4,500 customs and prohibition agents and a total of 11,000 men in the Coast Guard, who were mainly involved with lifesaving and aid to navigation maintenance. The coast was a sieve.

Canadian, Bahamian and Cuban ports had warehouses packed with contraband. The owners of hundreds of schooners, previously engaged in legal trade, were readied for work of another kind. Nassau Harbor, in the Bahamas, blossomed with the sails of smuggling vessels in numbers not seen since the block runners of the civil war. Their holds were packed with what was known as "hams", packages of six bottles, each wrapped with straw and sewn into burlap bags. The clientele were thirsty and ready to pay what the market demanded. Cargo schooners large and small put into inlets and out of the way harbors up and down the coasts. When eventually, Customs and the Coast Guard got better at interdicting liquor at dockside, the schooner captains would send the booze ashore in small fast power boats, or anchor in international waters and sell to the individual entrepreneurs who came out to buy from them.

Usually, unable to touch the mother ships in international waters, the Coast Guard concentrated of the fast contact boats when they made their dash for shore, and it was frustrating. They became known as (Carrie Nation's Navy). As time passed, both sides became more efficient and ruthless, as the Wets criticized the Coast Guard for being trigger-happy and the Dries berated them for not slowing the flow of booze at all costs. Never-the-less, in Coast Guard annals there are recorded many a successful chase and capture.

The Capture of the *Audrey B*.

One such incident involved the *Audrey B*, a 125-foot British flagged auxiliary schooner. The *Audrey B* was to deliver a large cargo of holiday booze to French and Spanish gang members off Long Island, the weather was too nasty for the contact boats, so the *Audrey* moved inshore. At 3:30 am Christmas day 1930, she was sighted running blacked out by the CG Cutter *PB290*. Chief Bosun in charge, had initially taken the *Audrey B* for a Grand Banks fishing schooner running in to spend Christmas with family, but when the cutter's crew hailed with season's greetings, she went hard over, bolting with full ahead engine and sail. The *PB290* pursued.

Convinced now that the *Audrey B* was the vessel that rumor had said would make a large Christmas delivery to the French and Spanish mobs, the Coast Guard was not about to let it slip away. On the foredeck of the cutter was a one-pound cannon, and the vessel also had aboard two 30-06 Springfield rifles, a Lewis gun and two Colt .45 automatic pistols – which should be enough persuasion if the crew of the schooner proved stubborn. Unfortunately, under sail and power the schooner proved dam fast. At 16 knots the cutter could keep up but wasn't gaining.

Repeatedly during the mist and fog of the pre-dawn hours, the *Audrey B* tacked, trying desperately to lose the Coast Guard cutter. Finally, she set course straight for Fort Pond Bay, northwest of Montauk Point. Doggedly pursuing, the *PB209* kept in the *Audrey B's* wake. On shore they could see a convoy of trucks and the fast cars drawn up, men ready to receive the contraband. They suspected she intended to run herself ashore and knew they must overhaul the schooner fast or lose her cargo. Finally, in hailing distance again, the Chief ordered the *Audrey B* to heave too but she held to her determined rush for shore. The cutter fired, and the *Audrey B's* stern disintegrated, as did her crew's motivation. Pulling alongside, the Coast Guardsmen found a crew of ten men and a cargo of 2800 burlap "hams" of six bottles each. Merry Christmas.

Smuggling has survived, but most of the big, beautiful schooners have not. Now the airplane, the motorship and the powerboat hold the day when it comes to contraband. Somehow, they lack the indefinable romance; nothing like the rush of a schooner slipping silently over the water on a dark night lit only by a scimitar moon. Lean back, close your eyes, see how they were; imagine them sail again.

Cargo Schooners, Tramps and Traders

Off Madeira's southeast coast, the fore top-sail schooner rolled and reared, yanked heavily back against her anchor chains. Over her bulwarks, men sweated crates of oranges from leaping shore boats. Two hundred tons – one crate at a time – and all on a gamble! The sun would be up soon, and Captain Hayes would see the *Susan Vittery* on her way while the fresh wind held.

Seventy-four days out of Fowey, the *Susan Vittery* was ready for her dash home. She had been to the Mediterranean first with coal; there she had loaded salt for Newfoundland. Twenty-one days out of Gibraltar, she hove under the lee of Quebec's Gaspe Peninsula and began to load cod while discharging the salt. For more than a week she'd plied the coast's obscure settlements, gathering her cargo, then in 18 days had run to the coast of Spain.

Loading oranges in a heavy surge.

A cargo for the Azores was found by luck, and Captain Hayes had it there and discharged before the great swells had begun to sweep Madeira's south coast. At Funchal a steamer had been forced to depart short of her cargo, and two fast fruit ships, steamers, lay at anchor, unable to load. Somewhere to the west, the remnants of a hurricane sent the great swells marching, and Captain Hayes would turn these to his employer's advantage. Ashore he had purchased oranges meant for the steamer at the wharf. If he could load and beat the fruit ships to market, a fine profit was to be had.

An hour past dawn, a consignment of wine was heaved aboard with some difficulty and much cursing. Six passengers and the clearance papers followed, and the schooner was on her way.

Out of Madeira's lee, the *Susan Vittery* made good time through the day. During the night a hard gale blew up, with seas crossing the swells. The schooner plunged ahead under flooded decks. On the midwatch, the lower topsail blew out and was cut away. At dawn a big sea broke over the port quarter, washing the ship's boat out of its gripes and tangled considerably. Passengers showed no inclination to breakfast that morning, but the weary crew ate heartily.

The gale moderated by evening and by noon the second day the swell had begun to ease. The

Susan Vittery reached northward, gaining time, for Captain Hayes knew the steamers would now be loading, and a race it would be for sure!

Hayes prayed for half a gale; he moved about the deck swiftly almost stealthily, keen eyes raised to the sky and the sails, alert to any advantage or change of weather. His manner was quiet but charged, and though he seldom raised his voice, there was an implicit obedience expected in his tone, which brought the best out of ship and crew.

Five and a half days out of Madeira, *Susan Vittery* reached the soundings with strong, steady westerlies, but 30 miles south of the Lizard, they encountered the worst gale of the season. At 9 p.m. the strong west wind went northwest and increased rapidly. All hands were called, and the upper topsails, course, and flying jib were taken in. Hayes, torn between the danger and the need for a fast passage, held what he could and drove the *Susan Vittery* up under the coast. A big sea was running, and the white-capped crests mounted higher and wilder by the minute. The air was so filled with spray that nothing could be seen of the bowsprit, let alone the land.

At midnight another reef was taken in fore, main, and mizzen; at the cost of sweat, bruises and torn nails. At last in the lee of the land the seas began to drop, even as the wind strengthened. Start Point Light loomed suddenly out of the gloom and the helm was put down a point. Eased so the *Susan* kept to it, every sheet flat, pounding, plunging up the channel at times driving her fine bow under to the fore hatch. Lee bulwarks long submerged, the *Susan* laid down, shrouds humming, driving into the teeth of the gale as only a schooner can. Not a soul on board slept, so exciting was the run. At 4:00 a.m. the Portland Lights were abeam, and by 7:00 a.m. St. Catherine's was but a mile north. By noon off Beachy Head under pure blue sky, the chop was driven into froth, so near the coast the waves had no room to build.

The wind came around to north by evening and eased. With Dungeness Light far astern, *Susan Vittery* tacked off the Deal and entered the Thames with the tide. Morning found her moored, with the cargo going to wagons. By fourteen hours, she had beaten the steamer and London paid well for oranges in temporary short supply.

Ten days later after refitting and supply, she was again underway for the Mediterranean and a salt cargo. Such was the world of the small cargo schooners.

Schooner Ram Victory Chimes

The greatly successful schooner evolved again and again as changes were called for by the special trades and routes she worked. In the coastal trade, the schooner came into her own. Less costly built than the brigs and brigantines of the same tonnage, and needing smaller crews, the schooner was efficiency itself to maintain and operate.

In Britain, the cargo schooners were generally two or three-masters, invariably with square fore topsails and seldom over 100 feet. North American schooners went another way. Although small, two and three-masters were built and used into the 20th century, it was the big three and four-masters that captured the trade. Handsome, economical ships four times the capacity of their British counterparts dominated the American coastal trade, particularly bulk cargoes, coal and lumber. As the type grew in tonnage, ocean passages became common. Some of the American schooners were giants, like the six-master *Wyoming*, second largest wooden ship ever built, and the *Rebecca Palmer*, first five-master to cross the Atlantic, and many others. Although they were sea-going, they were more competitive on coastal runs but did not do as well with the wind astern as the square-riggers.

A schooner too big

The seven-masted *Thomas W Lawson* was built for Captain John G. Crowley, as a double bottomed oil tanker, to be used for Pacific coast operations. Launched in 1902, her cargo capacity of 11,000 tons made her the largest sailing ship ever built by volume, outside of Zheng He's command ship, 1405 AD. There were several auxiliaries as large.

The seven mast Thomas W Lawson in ballast.

putting her to full draft. After a year as an East Coast collier, the *Lawson* was stripped of her upper masts and booms. Her wings clipped, she was towed as a sea-going oil tank barge, another attempt at profit.

In 1906, the Lawson was drydocked at Newport News and was re-rigged, with some additional sail area added at that time. The steel masts were now used to vent gas fumes while loading. They were named: 'fore, main, mizzen, spanker, jigger, driver, and pusher. With a liquid capacity of 60,000 bbl. she was chartered to Sun Oil for a Texas to East Coast tanker run.

By 1907, *Thomas W. Lawson* was under charter to a subsidiary of Standard Oil that decided to use her for the trans-Atlantic transport of oil. After loading Paraffin Oil at Marcus Hook on the

Delaware River, the *Thomas W Lawson* departed for London, England. Dropping down the bay in the last week of November she experienced fine weather. November is when North Atlantic weather tends to deteriorate and in the *Lawson's* case, for the next 20 days, it was purely nasty.

By the time she neared the British Isles, she had blown out most of her sails, suffered damages to hatches, deck gear and lost all but one of her lifeboats. Even the pumps were blocked with a slurry of seawater and coal. On crossing the Irish Sea as she approached the SW tip of England, her captain sailed too far north, and passed North of Bishop Rock Lighthouse. This caused the *Lawson* to find herself in the southern part of the ever-dangerous Isles of Scilly and uncertain of her position in the face of a building gale.

The captain anchored off Annet Island, trusting he could ride out the gale as was the practice on the American Seaboard. Recognizing the peril that the ship was in, lifeboats came out of both St. Agnes and St. Mary's, advising repeatedly that he abandon ship. The captain thought this ridiculous and declined, although he did request a pilot and a tug. A Trinity Pilot serving among the crew of the St. Agnes lifeboat, boarded at 5 p.m. on Friday 13 and he also suggested the *Lawson* be abandoned. As the weather worsened the lifeboats returned to their stations, the St Mary's boat with damage and one of her crew injured. From shore a cable was sent to Falmouth, some 60 miles away, requesting a tug, but the tug unable to make way in the teeth of the gale turned back to port.

During the night, the storm continued to strengthen even further, causing the *Lawson* to tack on her anchors, at around 1:30 A.M. her port anchor chain parted. A half-hour later the starboard carried away at the rim of the hawsepipe. Bow pivoting downwind, she drifted rapidly beam to the 30 ft seas, and drove starboard side too on Shag Rock. Raised and dropped, slammed down on underwater rocks by the tremendously heavy seas, the *Lawson* began to break up. Deck, swept by breakers, the crew scrambled up the masts to save themselves, but the masts snapped and tumbled into the raging froth on the ship's lee side. The stern was sheared off aft of mast six, drifting away from the capsized and sinking schooner.

From shore the Lawson's Lights had been visible until 3 A.M. and when they disappeared it was hoped that she had been able to set storm sail and slip her cables to sail clear of the rocks. At first light Sunday, it became evident that she had broken up and capsized. Commanded by Fredric C Hicks, the son of the pilot that had boarded the

Lawson, a rowing gig was launched to search for survivors. Only three were found, they had been swept through thick oil and surf a half mile to the Hellweather Rocks of Annet Island. The three were: captain, engineer, and a seaman who died soon after reaching shore. Unable to get a line through the breaking surf to the rocks, young Hicks, being the only swimmer aboard his gig, dove over the side without reluctance. Though encumbered by layers of

Thomas W Lawson deep loaded at anchor

clothing, oilskins and boots, he non-the-less swam the line ashore. Once there, he saw the injured men hauled aboard the lifeboat before he allowed himself hoisted back aboard. All this was done with a boat propelled only by oars in near hurricane force winds and water thick with oil from the wreck. They reached port as darkness fell. The body of Scillonian pilot, Hick's father was never found.

The silver medal of the British Lifeboat Institution was awarded Mr. Frederick C. Hicks for gallantry saving the captain of the *Thomas W Lawson*. He also, along with the crew of his gig, was awarded by the President of the United States: *for Courage and Humanity, a gold watch and chain for Fredrick E. Hicks, and gold medals for Osbert Hicks, William Trenary, Obadiah Hicks, Grenfell Legg, Frederick Hicks, William C. Mortimer, junior, and Israel Hicks, junior, boatmen, of St Agnes, which have been awarded to them by the President of the United States in recognition of their services in rescuing the survivors of the crew of the American schooner Thomas W. Lawson of Boston, which was wrecked off the Island of Annet, Scilly on December 14 1907.*

The islands and rocks for a hundred miles, were coated with oil. The wreck of the *Thomas W Lawson*, besides being a tragedy all involved, it was the first great oil spill.

The coastal trade

By the 1930's the price of fuel had dropped so low, that steamers gained a real competitive edge over the big schooners. This fact, coupled with the depression, led to the disappearance of the larger schooners, many turned into barges.

Three mast fore-topsail Schooner

Loading Cyprus at Orange Texas

The small schooner had a place in every ocean it the world though; they could make a profit in visiting the little places that offered a larger ship nothing but lost time. In the thousands of atolls of the Pacific, from Australia to the Straits of Malacca, they were the lifeline connecting forgotten places with the world.

European schooners found themselves a niche, near home. For the British and Scandinavian schooners, this was particularly true. In the home trade they visited small ports and islands most of the year. At some point they would load for the Mediterranean, discharge and take on salt and then sail for Newfoundland or Labrador. The schooners would stop at a hundred little cuts in the rocky coast, loading dried or salted cod and discharging salt. With a load aboard, the schooners would cast off for England or the Continent with every stitch set, being driven like yachts for the best of the new market.

It was World War II which finally brought centuries of trade under sail to a close. The war brought about such a development in industry and transportation that the little ships had nowhere left. A few managed to tramp the Pacific and Caribbean but by 1955 the cargo schooners, the last of the centuries of commercial sail, had passed from all but memory.

Masefield put it most eloquently in verse:

"The schooners and their merry crews
Were laid away to rest
A little south of sunset
In the islands of the blessed."

The Cargo Ships of the Indian Ocean

The Dhows

Dhow is a European word taken from the Swahili name for a boat, (*daw*). Ship in Arabic is *markab* and *safiinah* and in the Quran, *Fulk* is used for many types of cargo vessels. African and Indian mariners in the Indian ocean and its connecting seas and gulfs. Versions would have been as common in ancient Egypt as they are in modern Egypt where *Gaiassas* also called *Feluccas,* still ply the Nile and canals.

The types are many, and every region of the Indian Ocean has developed a type of vessel to fit its needs but the history and development of very few are known. Although, Europeans name all boats of the Indian Ocean Dhow.

In the Middle East however, boats are classified according the shape of their hull, thus, dhows with square sterns are described as: ganja, sanbuuq, gaghalah and jihaazi. The square stern was a European influence, only seen after the arrival of the Portuguese in 1500 AD. The traditional form of construction of double ended is found in ships now called zaaruuq, buum, boom, badan, etc., and still have the double-ended hulls that come to a point at both the bow and the stern.

The Felucca

These inland or river cargo barges are used from Egypt to India.

Types are many: Baghla, Sambuk, Boom, Ghanja, Zaruk, Badan or Indian dhows like the Doni, Pattamar, and Mashwa. The early dhow was known for two characteristics. First, it's quadrilateral or semi triangular lateen sail, and secondly, for its stitched construction, which involved the ancient practice drilling holes and sewing the planks together with rope or twine cords of various fibers.

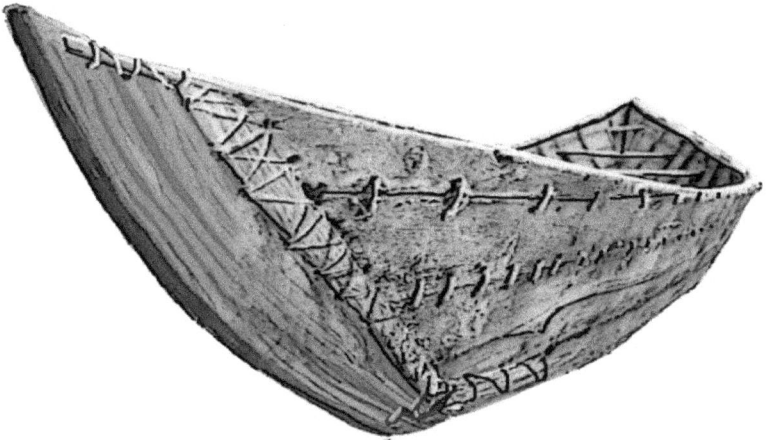

Sewn Hulls

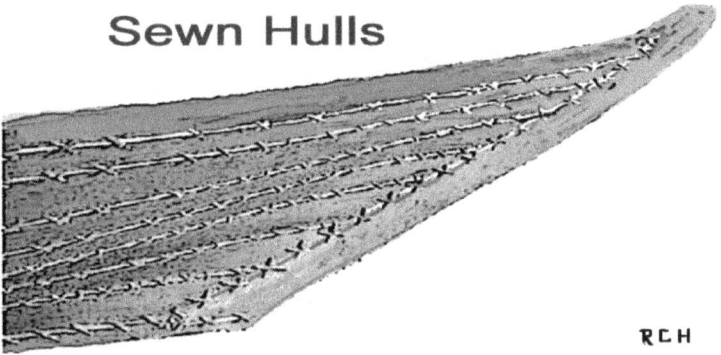

RCH

In many regions of our world, stitched hulls are still common. In the Indian Ocean, it was the major technique used to hold ships together and was the norm before the fifteenth century arrival of the Portuguese. In the Mediterranean, marine archeologists have found remains of sewn boats that date from the early Minoan and Egypt era, to the Roman Imperial age but mostly European shipbuilding technology took a different path.

A 13th century a Venetian merchant gave his opinion of the sewn vessels, then navigating the Persian Gulf. He described the ships as of the being the worst sort, being hazardous in their use for voyaging, and endangering the traders, risking both life and goods. Experienced in the Mediterranean style of shipbuilding, he reported that although these ships were watertight and enduring, they tended to come apart in storms.

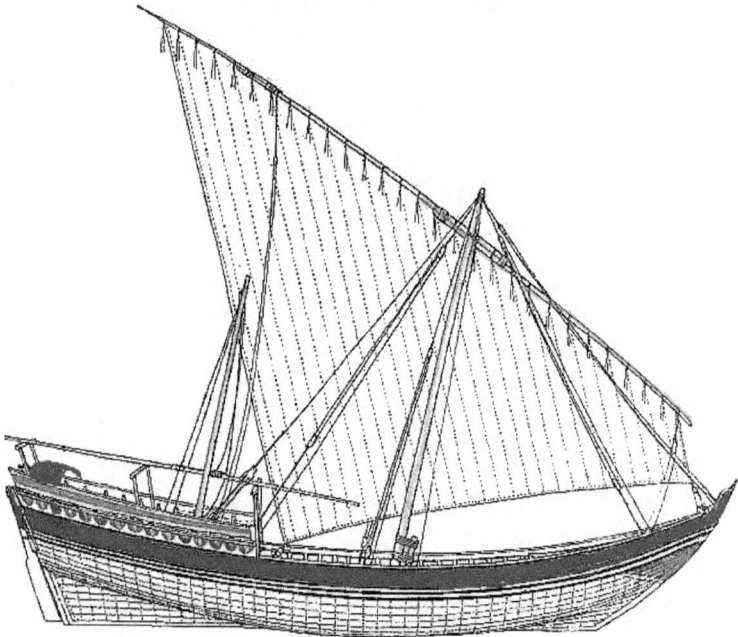

Red Sea Dhow shares much with the Carvel.

None-the-less, ships stitched together in this fashion were the norm 5,000 years ago. They endure and can still be found today on Sir Lankan coast and on Africa's Swahili Coast.

Seams were caulked with cotton and tar or Bitumen. Marine parasites were held in check with sulfuric powders mixed with fat, oils or tars; sometimes thin sacrificial layers of wood were nailed on. The original sails were made by weaving palm leaves and fibers but eventually cotton cloth became the preferred fabric for ocean voyages. Sails were made of various cloths, over-lain and stitched parallel to luff and leech. Sails were cut differently, flat for windward and curved for downwind use and with no method of reefing sail, they carried smaller sails for use in high winds or storms. In some rigs, it's easy to spot European influence. The Sambuk of the Red Sea is a twin to the Carvels, so popular in the Western Mediterranean Sea. Only a slight difference in the bow and the European preference for the square sail distinguishes them.

Sambuk

The Red Sea Sambuk's hull
has a great deal in common with
the carvel, (Columbus's Nina
and Pinta). Even the stern and
rudder are alike. Only the
rigging differs.

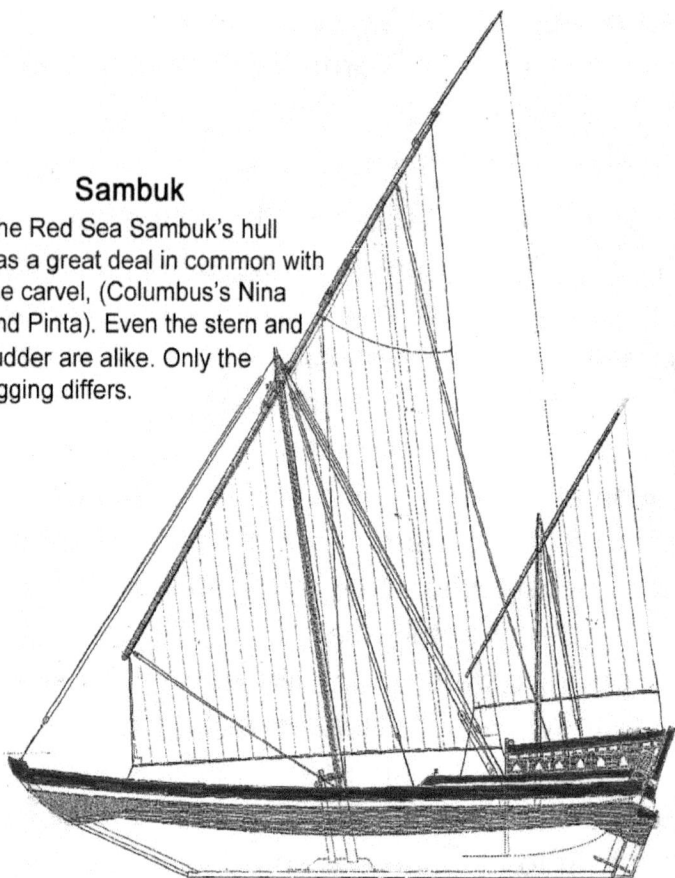

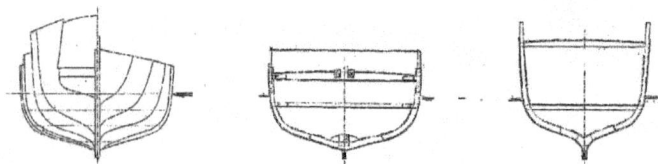

The Baghla is a ocean going ship that was enfluenced by Portuguese ships of 1600s.

Baghla

The big Baghlas have sterns that are much like the armed Portuguese trading ships that terrorized the coasts of the Indian ocean during the 16th century and called it just business. Like the Sambuk though, the bow of the Baghla was fine, far more rakish the European Galleons. These ships were primarily cargo carriers and were seldom heavily armed. They are still built on the coast of Kuwait. Baghlas are ocean going ships, that sail to Madagascar, Bombay, even Calcutta but always in accordance with the seasonal Monsoon winds at their back. A dhow captain would tell you, *"only idiots and Christians sail to windward."* To this day, these vessels are moored three and four abreast at the docks of Arab and Indian ports—a conspicuous contrast to giant steel ships and modern skyscrapers. Though most have been stripped of rigging in lieu of diesel engines, the dhows' construction is otherwise unchanged. Cargo, ranging from sacks of grain to all sorts of general cargo, is loaded and unloaded the traditional manner, using manpower.

The Boom

The Boom was very common to the Arabian Gulf, along with the larger three mast versions called Ganhjas. You don't see them anymore, except in the racing vessels of the region. Dubai's rulers came up with a plan to retain the traditional building skills and the lore of dhow sailing through competition, by building and racing dhows.

The name of the modern regattas is the *al-Shandagha*, named for the point of land extending out from the entrance of the Dubai Creek which is landfall for all vessels arriving at Dubai's port. It is the first of a three-race series, sailed by a completive fleet of boats 60-foot in length. Like the out-island races of the Bahamas, the boats lay at anchor until the signal to start, then all sail is set, and the anchors are raised with muscle power, only blocks and tackle are allowed to sail the boats. These wooden boats must be traditional in the style and construction, all like the coastal merchant vessels that have plied the trade routes between Arabia, India and East Africa for centuries.

On one of my first international voyages as a young man, my ship arrived in Mombasa, Kenya with general cargo out of Bombay. We were to lay at anchor for two weeks before our dock and a warehouse were available. Ship was more laid back in those days. Cursed with youth and the spirit of adventure, I talked a fellow seaman into standing my watches (with the captain's permission) for a two-week period. I then departed the ship to play tourist. A day in Mombasa was enough but I'd always been interested in Zanzibar. It was December, with the northeast monsoon blowing and the Sultan's red flag was pressed up against the mizzen rigging. I boarded the seventy-foot combination cargo, passenger vessel with the optimism of youth. It was adventure; I was bound south to Zanzibar Town.

It was my first experience aboard a Dhow, and a different and amazing two-day passage. The deck was cluttered; smelly would be a compliment. The sanitary facilities amounted to a box protruding over the stern with an oval hole. A shower weas easy, a bucket of sea water. They had a communal water keg with a dipper, and meals were rice with a little something different mixed in each day. Food it was but to my spoiled pallet, nasty food. What I'd describe as cream of wheat gruel was the morning meal, nuts, dates and fruit any old time. Accommodations, there weren't any; (you curled up where-ever). We had some calms and some good breezes. Boy did that old tub fly when the wind picked up. She had a mixed black and Arab crew and they were real old-time sailors. The main yard of the lateen sail was 100-foot-long. If it blew too hard, the plan was to lower the yard and bend on a smaller sail; I doubt that would have ever been done.

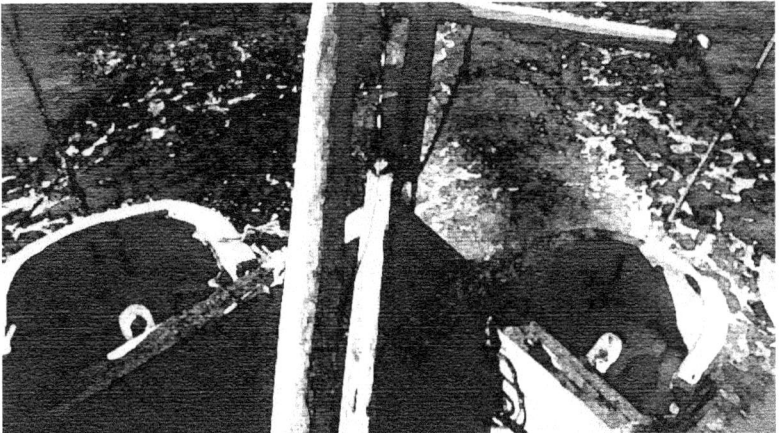

Thunderboxs at the stern of the dhow.

Dhow with bowsprite, even today common to East African coast.

These guys would have sailed her under first. All day on deck you roasted, and at night you slept wet but it was worth it. The experience on that dhow was one that has stayed with me down the years.

Dhows at anchored off Zanzabar

We came in past what I'd estimate, were 50 big dhows at anchor. The crew dropped the hook just north of the port and I was rowed to shore, where at the customs house, I presented myself and my passport. I'd planned a day or two to see the sights; I would then catch the ferry, 40 miles to Dar es Salaam, Tanzania. On a rented bicycle I peddled around the island and visited the shops and spent money on neat stuff. The Island was a Sultanate then, and there was a lot of friction between the Arab Omani population and Black Africans----the kind over which men slit throats in the dark. White boys weren't overly popular in some quarters either---yet, on the surface all was peachy peaceful. At the time, Britain was devesting itself of colonial possessions at a prodigious rate. The tribal and racial politics were a bit passionate for a young lad from the states.

Three days later, I was on a dilapidated bus. It was packed inside and on the roof. It headed west first, and then north to Tanga, before crossing the border into Kenya and on to Mombasa. Kenya had its own tribal problems, to mention one, the Somalis wanted the border moved south so they would be part of Somalia rather than Kenya well that didn't

Landing in a stitched shore boat. These liter cargo ashore.

happen. That whole bus ride is another story. I was back aboard my ship five days early, and not much tempted me to wander around on shore after my foray---that and I was out of money anyway.

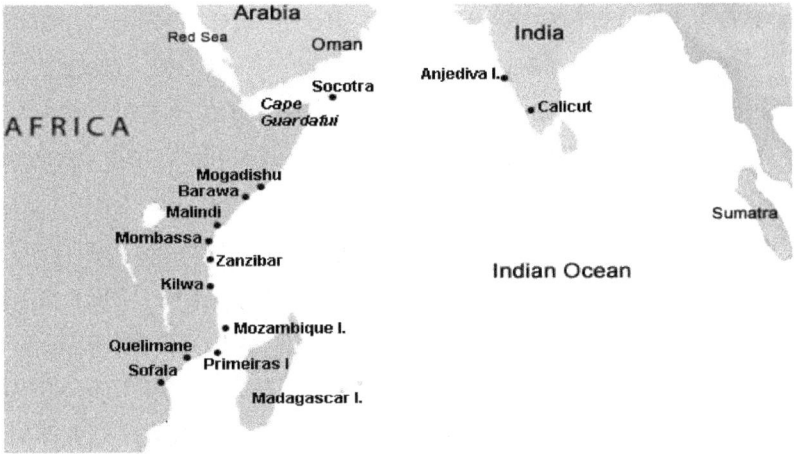

The coastal area I had sailed down was merely a short hop for a Dhow. I'd covered only 270 miles in the center of what's called the Swahili Coast, which stretches along Africa's east coast from the from the Red Sea, to South of Madagascar. Migration and trade on this coast had been accomplished for millennia with the use of big dugout canoes, and eventually small sailing vessels. The Bantu were the pioneers of this region and their coastal voyages was helped by long stretches of islands and coral reef that presented a barrier to the open ocean. It was a calm highway along the coastline. The many river estuaries of the coast of East Africa offered exceptional natural harbors, access to the interior and inland markets.

At the end of the 7th century AD, Arab trading bases established on the coast began to draw even more black Africans from the fertile inland regions for trade. From the mid-8th century, Muslim merchants from Arabia and Egypt were creating settlement towns on the Swahili coast and making its islands their homes. In the 12th century AD, Persians, known as Shirazis' arrived, Arabs, Persians intermarried with each other and with the indigenous Bantu. As the races mixed, so did their languages, which resulted in the blending of cultural, the evolution of the unique Swahili culture. **Swahili** means literally, (**people of the coast**).

The trade settlements expanded into independent coastal trading cities, north to south: Mogadishu, Mombasa, Zanzibar and Kilwa. Trade goods came down rivers from distant inland realms, to places where regular markets were established, cities like Kilwa, Sofala of the kingdom of Zimbabwe. At the pinnacle of their success, a period which spanned the 12th through 15th century AD, merchant city-states traded with African tribes living as far inland as the Great Riff Valley and Lake Kariba in the southwest. By sea, they were connected across the Indian Ocean, with India and Sri Lanka, even China and Southeast Asia. Sea voyages across the vast Indian Ocean were synchronized with the rotation of monsoon winds which blew to the northeast in the summer months and which alternated to southwest in the winter months. Comparatively, sea travel was swift when compared with land travel.

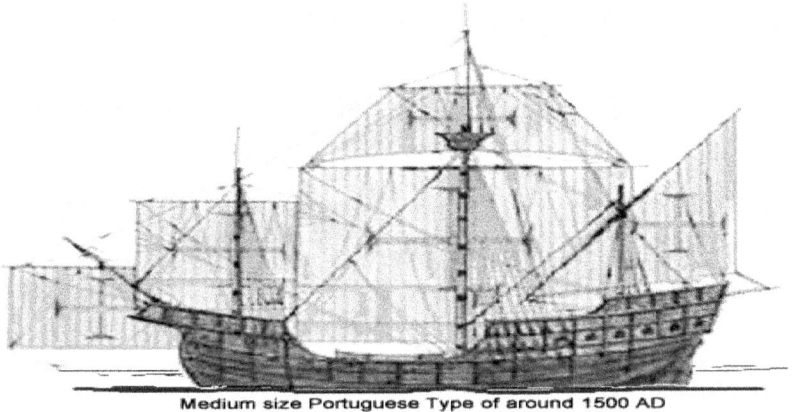

Medium size Portuguese Type of around 1500 AD

The rounding of the Cape of Good Hope by Vasco da Gama in 1499 AD sounded the death knell of the Swahili city-states. The Portuguese mariners that followed da Gama sailed up the east coast of Africa, seeking full control of the Indian Ocean's system of trade. Dhows were not warships and became easy prey. The disorganized trading cities had little chance against the tactics and superior weapons of the Portuguese. The Swahili Coast was not able to defend itself. Strategically the Portuguese built trading forts to control the entire Indian Ocean. They had a major military base at Goa, India, and others in Africa like Sofala, Mozambique Island and Shama, plus many small strongholds. Rather than establishing fair trade, they simply took what they wanted from the Swahili coast, sacking cities and burning what they couldn't carry away. Expeditions up rivers into the inland African kingdoms, were neither to trade or administer, but to remove anything of worth, and to do so with minimal cost. Rivals were destroyed or sunk on sight. When the Portuguese finally pulled back, it was not because of any resistance put up by the Swahili but the terrible mortality of tropical diseases that discouraged the Europeans'. They eventually deserted the central coast and focused on the coast south of Madagascar. They went after the gold of the Mutapa Kingdom of inland Zimbabwe. The disruption and destruction wrought by the Portuguese on the Swahili Coast had broken the age-old system. The region went into permanent decline. A few cities managed to reestablish some trade but only after the Omani Empire enveloped them. Unfortunately, the Omanis were primarily interested in ivory and slaves.

Eventually the British established colonial control, and after a few centuries their efforts curtailed the slave trade. The semi peaceful period ended with the post WWII departure from the colonial system of European nations. From what little I saw; decolonization created an unpleasant upheaval for the various populations of the region. As things happen, it was only a year a year later when the Africans of Zanzibar revolted and slaughtered the Arab population of the island. Many were driven out to sea on overloaded and under provisioned dhows. On shore, the numbers of dead were estimated as high as 20,000. Numbers depend on who you're listening to. At sea---well, a lot of those big dhows were never heard from again. That was the beginning of the end for dhows. On the date of the revolution, I was getting in other troubles of my own, and doing it on another continent. I heard nothing of Zanzibar's troubles for years. We didn't have 24 hour a day international news back then. What I do remember is the rush of wind and water---ahh----the memory of those two days and nights under a billowing hard-pressed lateen sail!!!!

The Vessels of the Pacific Ocean

Junks, Djongs, Proas, Double and Outrigger Canoes.

After thousands of years, the 100 foot long Jaluit Proa was still carrying passengers and cargo durring WWII.

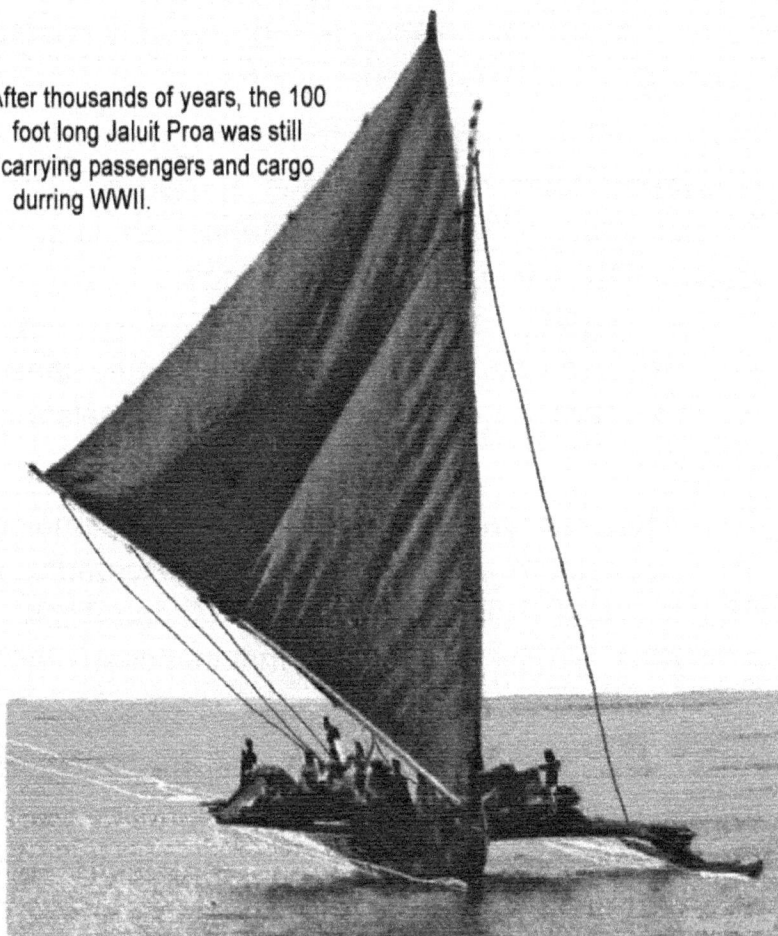

Studies show evidence that humans began migrating to Australia and the islands Southeast Asia before 65,000 years ago. The sea level was far lower at that time. Following the coasts and island hopping would have been easier. These populations shared many physical traits with North Africans, but strangely in 2019 genetic studies found that Austronesians are more closely related to European and Asian populations.

Most scholars agree that the core group of the Neolithic people originated and spread out from Taiwan, but with time the tribal cultures of Austronesia varied widely from region to region. Some 1,257 languages share the same Austronesian linguistic foundation. All these regions share the same domesticated animals and plants.

This group spread out over an area extending from Taiwan to Australia, East to Micronesia's Easter Island (Rapa Nui), west to Madagascar, a vast area of the Pacific and Indian ocean, their coasts and islands.

Austronesians existed primarily in the world of islands, migrating to and settling in regions with tropical or subtropical climates where there was abundant fresh water and fertile land. They never seemed interested in moving inland, not on the mainland or even on the large islands.

70 foot Double Canoe with matted claw sails.

The early Austronesian peoples considered the sea to be the foundation of their lives. Regardless of their roots they developed into seafarers. They built seagoing vessels large and small, differing in shape and construction but all of which navigated over a vast area. We know the sorts of vessels they sailed in ancient times for many remain on remote coasts and archipelagos.

300 BC, a 100 foot tanja rigged proa capable of a 60 ton cargo.

Double canoe hull, catamarans, evolved to asymmetrical double hulled proas. Outrigger canoes morphed to trimarans and of course, mono hulled vessels. The sails and rigging of these vessels were some of the most innovative and efficient on earth. Bipod or tripod masts, rigs like *Crab Claw*, and the *Tanja* rig which evolved into versions of the lateen rig favored in Arabia. Of course, the Malay and Chinese battened lug rig. Customarily the sails were made from plaited mats of the tough and salt-resistant pandanus and palm fronds. The influence of the Austronesian peoples is everywhere though; they are like ghosts in history, always building in ways that improved their ability to be voyagers.

Mainland China

The population of the mainland of China were mostly river people until around the end of the **Tang Dynasty**. Seafaring and trade routes did not exist for them in a large degree. Ships were more often sailed by the peoples of other lands. Before 4500 years ago the mariners navigating the coasts of east Asia were mostly the Austronesian sea peoples.

300 BC to present, tanja rig cargo carrier.   This three mast ship would be about 110' in length. Indonisia has many simular types and sizes. Caracor is one. Versions are found even in Madagascgar.

mariners navigating the coasts of east Asia were mostly the Austronesian sea peoples. Scholars of the **Zhou Dynasty** (1047 BC to 247 BC) wrote of huge Austronesians ships that traded with Chinese ports. They were termed *Kunlun po,* (transport ships of the dark-skinned people). They are recorded again in around 250 AD and described as capable of carrying 800 tons of cargo, 700 passengers, being high-sided and hundred-fifty feet in length. They're mentioned as being hired to transport large groups of Buddhist pilgrims to India.

The rig of these ships was described as having multiple staggered masts, (as many as six) the sails of each interacting with the other to create a venturi effect or cause each sail to reflect the force of the wind into the next. The rig was said to be low and not influenced adversely by storm winds. Using strong winds efficiently, they drove the ships at very good speed in all weather. ---- **Hmm!** ---- This multiple mast, staggered rig reminds me of the great 15th century AD ships of Zheng He.

Chinese Junks

There have a great many types of junks used over the thousands of years of civilization from the north to South China Sea, and neighboring regions. Original concepts and adaptations of the ideas that arrived with ships from the west. Many of the things the Chinese came up with were excellent. Things like water-tight bulkheads, compartments open to the sea to dampen the motion in rough water and even the use of a single strong rudder at the center of the stern. The fully battened Chinese Lug rig was a brilliant innovation that controls the shape of the sail and shortens area by simply lowering it by sections. There are other things that western seamen simply shake their heads over.

There were just as many parallel inventions like ballast tanks, what westerners once called wet wells, for keeping fish alive or diminish the toil in ballasting when a ship had no cargo to stabilize it. These features were found in Roman ships as well as Chinese and of course, ships had the ability to pump the water back out when necessary. They went about it with very different designs, but it amounted to the same thing. The floating compass was another but this one though improved to perfection in the west, while in China it remained virtuously unchanged and inaccurate for a thousand years. Of course, fog and nasty weather were more prevalent in Europe. Perhaps the Chinese weren't motivated.

Most junk types were flat bottomed, and some did use a form of dagger-board to reduce leeway, usually in the forward part of the hull to keep the center part of the holds clear of obstruction. This threw the balance off and led to the development of the balanced rudder, a feature common to many junks. I was once told that a big junk could require a dozen men to steer in strong weather, perhaps one of the reasons holes were cut all over the rudders. (To relieve pressure)?

Song Dynasty Junks

Seven hundred years later, 10th to 13th century AD, the **Song Dynasty** was ruling China. The Song were pro-trade and began building large ships. These were modeled after the deeper draft, v-bottomed hulls of Austronesian ships of Southeast Asia, rather than traditional flat-bottomed vessels of China's rivers. These were sea going ships. Most had three masts with Chinese lug rig sails and were on average, over a hundred feet in length. They were also beamy and drew around seven feet of water. All had the usual lifting system to change the depth of the rudder. They had around ten transverse watertight bulkheads and used many of the shipbuilding technologies and materials of Indochina and southern Thailand as well as Chinese. These ships made regular trading voyages to Indianan Ocean ports and sailed south throughout the island ports of Indonesia.

10th to 13th Century Song Dynisty, Quanzhou Ship.
110' length and 30' beam

The Mongols invaded China in 1234 but the southern Song maintained their rule for another 45 years in the south. In 1279, *Kublai Khan*, grandson of *Genghis Khan* founded the **Yuan Dynasty** after taking the last Song stronghold. The fleet was not only important to the Mongols for trade, but for war. They used the ships for invasions of both Japan and Java though neither campaign succeeded.

The 14th century Yuan ships arrived in the western Indian Ocean with each season's monsoon. This was the middle-ages in Europe was still experiencing problems with Mongols, and Tartars but Europeans were also benefiting from open trade routes with the East. Marco Polo detailed the construction of Chinese ships and their huge size in the account of his travels. The enormous dimensions of the Chinese ships of the Medieval period, as described by Chinese sources, are confirmed by other travelers to the East. Telling of his 1347 voyage from India to China, *Ibn Battuta* wrote:

Sailing east, my party arrived in the port of Calicut and disembarked. There were at the time thirteen Chinese vessels, preparing to depart. On the seas of China, traveling is done only in Chinese ships, so we shall describe their arrangements. The Chinese vessels are of three kinds; large ships called chunks (junks), middle sized ones called zaws (dhows) and the small ones kakams. The large ships have anything from twelve down to three sails, which are made of bamboo plaited into mats. They are never lowered but turned according to the direction of the wind; at anchor they are left floating in the wind. A ship carries a complement of a thousand men, six-hundred of whom are sailors and four hundred men-at-arms, including archers, men with shields and crossbows, who throw naphtha. Three smaller ones, the "half", the "third" and the "quarter", accompany each large vessel. These vessels are built in the towns of today's Quanzhou and Sin-Kalan. The ship has four decks and contains rooms, cabins, and saloons for merchants; a cabin has chambers and a lavatory and can be locked by its occupants. This is the manner after which they are made; two (parallel) walls of very thick wood (planking) are raised and across the space between them are placed very thick planks (the bulkheads) secured longitudinally and transversely by means of large nails, each three ells in length. When these walls have thus been built, the lower deck is fitted in and the ship is launched before the upper works are finished.

Translated from the writings of, *(Ibn Battuta).*

The peasant's rebellion of 1368 AD ushered in the **Ming Dynasty** that lasted until 1644 AD. There was little change in naval or trade policy in the early decades of the Ming's rule.

The pinnacle of Chinese Maritime Accomplishment was reached under the rule of Zhu Di the Yongle Emperor, who ascended the throne in 1404. Zhu Di accomplished this through an extraordinary slave, *Zheng He*.

Captured as a boy, Zheng He was brought to Nanjing. Like all slaves being entered into imperial service he was castrated and entered into training for imperial service. He was then sent to Beiping (present-day Beijing) to serve in the palace of the emperor's forth son, Zhu Di, the Prince of Yan.

Zheng's intelligence and devotion won the trust of Zhu Di's and he was chosen to serve as the prince's personal guardian. Zheng He's intellect and leadership skills became apparent. Zheng He fought at the side of Prince Zhu Di, for several years, on campaigns and battles throughout China and was influential in Zhu Di's securing of imperial power. As the *Yongle Emperor*, Zhu Di promoted the eunuch official Zheng He, in 1403 AD, elevating him to the position of *Grand Three Treasures Eunuch.*

10th to 14th century, the Song Dynasty built bigger deeper ships. In the late 13th century the mongols needed them for invasions of Japan and Sumatra, but it was the Yaun's Zheng He, who built truly big.

Toward the end of 1403 AD, Zhu Di assigned Zheng He to study the subject of maritime affairs. Zheng He first directed and comprehensive examination of nautical charts, celestial navigation, almanacs, astronomy and the marine sciences of shipbuilding and maintenance. He was then instructed to oversee the building of the **Great Treasure Fleet**. Ships large and small amounted to a fleet of 200 vessels. Zheng He made seven voyages between 1405 and 1443. The fleet sailed from the South China Sea to the Indian Ocean into the Persian Gulf and Red Sea, and the coast of Africa. It has been theorized that they also reached the Americas and sailed around Australia. These were a combination of trade, diplomatic and exploration missions and although they surveyed and charted a large part of the world, the new emperor ordered much of the fleet destroyed and all the records burned after Zheng He's death in 1443. Shipbuilding skills gradually diminished in following years and the navy fell into disrepair. For a time, China focused mainly inward.

As for the Great Treasure Fleet's record of accomplishment, the Admiral had stone markers placed on far continents during voyages and Zheng He had commissioned a monolith with a record of his exploits chiseled into it. It was placed in the garden of a remote monastery and remained unnoticed for six centuries. Some charts were reported to have reached Europe.

There is even a story of Magellan having a copy of a Chinese chart showing him the straits that now bare his name?????

1405 AD, Zheng He's 420 foot Treasure Fleet Command Ship.

A list of the ship types making up Zeng He's treasure Fleet "from Lou Maodeng Book"

Command ships used by Admirals of the fleet and their officers. These were nine masted junks, claimed by the Ming Shi to be about 420 feet long and 180 feet beam.

Horse ships, carrying tribute goods and repair material for the fleet. (eight masted junks, 340 feet in length to one-hundred-forty feet beam).

Supply ships, carrying provisions and stock items for the crews of all ships. (Seven masted junks, 260 length to 115 beam.)

Troop transports, (six masted junks, 220 feet in length by 83-foot beam.)

Warships, (five masted junks, 165 feet long.)

Patrol ships, (8 oars to a side and 20 feet in length.)

Water ships, (each carrying a month's supply of water for the fleet.)

The number of men and the quantity of supplies these ships, this fleet required---it was staggering. They carried everything from blacksmiths to courtesans, and accommodations for the upper ranks were plush.

During the isolationist period of the new Ming regime, (it extended from the mid-15th to early 16th century) all seaborn trading outside of China's own waters was forbidden. With the navy in decline, piracy became a problem, particularly Japanese and Taiwanese pirates. The ships of bordering kingdoms stepped into the void controlling trade and expanding their ships in size and numbers. By the end of the 15th century, the armed merchant ships of the Portuguese, Dutch and Spanish were also in play. By the mid 1600's the Ming were at war with the Dutch on Taiwan. So, it goes.

The Java Djong

The Javanese ship, the *Djong* was built primarily in areas of north Java and on the south coast of

Borneo and nearby islands, in area with large teak forest. Teak was a favored material because it was somewhat resistant to the **teredo navalis** or ship worm as well as other marine parasites. Even, to this day, large sailing cargo vessels are being built on these coasts, though now, most have fabric sails and diesel engines.

15th to 17th Century Javanese Djong Ships were often over 300 feet in length.

The difference between the Javanese junk (*called a Jong*) and a Chinese junk was vast. The Jong was also double ended but carried two oar-like rudders and was generally rigged with Tanja sails, although the Malay junk sail was used at times. The Javanese ships were built with layers of thick teak planks while the Chinese used light soft wood. As a Jong aged, planks were added. The lines and sails were of plaited rattan that was tough and lasted for years. The Jong's hull is formed by attaching planks to the keel and then to each other with trunels and adding the interior framing or reinforcement afterward. Holes were drilled in the planks from the inside and the trunels pounded in They are not visible on the outside of the hull. Smaller boats were often stitched together like Indian Ocean Dhows. The vessel was similarly pointed at both ends and carried two oar-like rudders like Mediterranean Galleys. The Chinese junks on the other hand used iron fasteners, had watertight bulkheads and a mounted a single rudder affixed to the transom.

Jongs were large, often stepping six or seven masts and were usually built and sailed from Indonesia's Muslim Sultanates. Europeans who engaged them in sea battles came off on the short side for they were impervious to cannon fire. After two painful occasions, in the year 1513 Dutch ships avoided them. A Venetian mariner described them as having the strength of a castle wall, for the multiple layers of teak planking was invulnerable to shot of the time. The Chinese forbade them from entering port for one Jong could defeat a dozen war junks.

Asian trade from 1600 AD thru 1900 AD

16th century southern Junk   120' long and 30' wide
They had a deeper draft shaped like western ships

Chinese merchant junks were prevalent in Asia from the 16th century up to WWI. They competed with European ships of all sort as well as junks of the Japanese, Malay, and vessels of Indonesia and the Philippines. The Long-

range ocean-going junks were fairly large. These ships had at least three masts and had a compacity of between 250 and 900 tons. Crews were up to 140, plus merchants and passengers. A total of 300 was not unusual. Big trading junks visited Europe and US ports on both the east and west coast. Although there were many big junks, most were small and medium size with a compacity of from 50 to 200 ton. They could slip into small ports and sail up rivers as well as make coastal runs. Often, they would liter cargo ashore or transship a portion of a big ship's cargo. By the 1800s, many had the mixed features of eastern and western vessels.

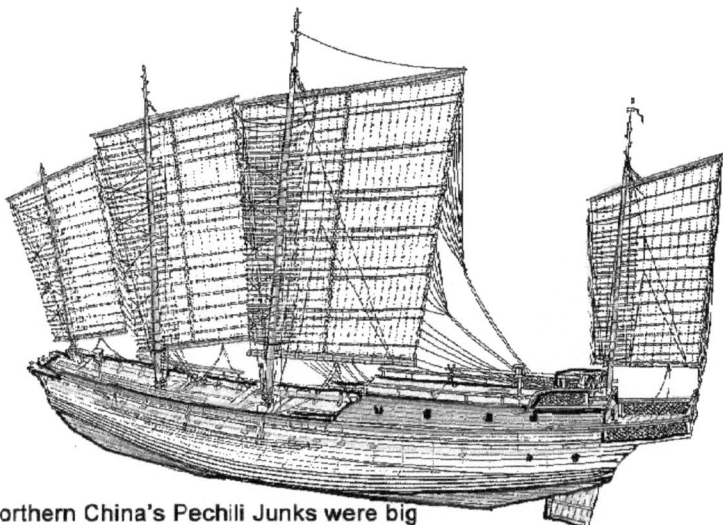

Northern China's Pechili Junks were big
often up to 200 foot long but less than half the size of those in 1400 AD.

Today, junks are still a part of the scene. They have engines now and the rigging is used primarily for hoisting cargo. The vast rafts of junks, manned by families who live aboard and make their living from the vessel, are thinned out more with each passing year. I expect that soon, if you want to experience a Junk, you'll have to visit a museum to look at pieces and models. Personally, I love those life size replicas.

Junk common to South China with obvious western influance.

Junk common to the Gulf of Thailand

The Iron Windjammer

The Aging Victor

It was a clear, windy night in the South Atlantic, the seas marching west before the full strength of the Tradewinds. The year was 1934; and the PNO liner was bound for Rio. On board, the last passengers had left the ballrooms, and the Stewart's crew scurried about their late-night cleanings and services. On the bridge, a lookout advised the young third officer of a distance sail off the starboard bow. At 0400 the second officer arrived for his watch; the distance between the vessels had halved.

Herzogin Cecile

At dawn the Capt. arrived on deck and took the mates glasses. The liner was gaining but slowly now and what dawn had revealed was a lofty, four masted bark going well under lower, gallants. A magnificent sight seldom encountered by the passengers, like the ones watching at the liner's rail! Well into their voyage, the passengers could use a bit of novelty, the captain decided. He would give them a sight to remember, by steaming full ahead and crossing the bow of the elegant but antiquated sailing vessel. The order was passed, and on the telegraph, the mate rang the engine room for full power.

Aboard the four mast barque, *Herzogin Cecile*, the Capt. arrived on deck after his

breakfast; surveying his ship, he inquired about the liner overtaking on his port quarter. The mate informed him of the liner's increased speed since dawn. Using his glasses Capt. Sven Erickson could see passengers packing the liners starboard rail. Well, if it was a show the British wanted, it was a show they would get. Erickson ordered all-hands- on-deck to set the topgallants. Able-bodied seamen and apprentices scrambled into the rigging. As the steamship drew up on the barque's beam, closing to 300 yards, the *Cecile* matched her speed, now under the press of new canvas. Ships of past and future ran neck and neck with the throb of engines straining to out-match the power of wind pressed canvas. They raced side by side, the pride of two eras at stake on an empty sea.

"More power!" The captain of the liner ordered. Give me another knot!"

His grumbling chief engineer ordered more stokers roused and put on the line. The barque' s Captain, also called for more power, and the Royals fell from their gaskets to be grasped by the wind, now blowing half a gale. Below, in the liner's bowls, men shoveled like demons panting in the heat, while the Windjammer's crew stood waist deep in the cold water, furiously cranking on the powerful deck winches to trim sail.

The steamer had begun to inch ahead but the old sailor matched her. Lee rail at the water's edge, the old barque tore through the seas like a flung harpoon, bow wave spawning rainbows in the morning sun. Borne on wings of canvas, the *Herzogin Cecile* began to pull away. Faster and faster 19, 20, 21 knots as the PNO liner fell slowly astern.

The passenger liner gracefully, incredulously, dipped it's ensign and crossed the barque' s stern, saluting with three blasts on the foghorn. Defeated by beauty and spirit, a thing that lived only in the wind driven ships, the liner set course again for Rio. The *Herzogin Cecile* dipped her finish flag of blue-and-white and held towards Australia.

The big Iron-hulled sailing ships

represented the final development of commerce under sail. They were designed to carry passengers and bulk cargo for long distances as the switch to riveted iron construction began taking hold in the 1870s. The early ships were heavy for their size and were notorious for shipping seas on deck. There were numerous complaints of mad captains driving their Australian immigrant ships across the roaring 40s with decks submerging and terrified passengers battened below for weeks on end. One early iron three master, an Emigrant ship of about 270 feet had log entries that read something like this.

Three mast, ship rigged vessel clearing the sea buoy, sailing close hauled.

SHIP'S LOG: *In the Bay of Biscay, at 10 a.m. the wind freshened, all hands called to shorten, topgallant sails, crossjack, spanker, and outer jib were secured in strong squalls and heavy rain. During the night a full gale blew hard putting up a lofty cross sea, through which the ship plunged with her decks flooded fore and aft. At 2 a.m. the main topgallant backstay carried away, requiring heroic effort to save that upper mast. At 8 a.m. a rogue sea rose up and broke over the starboard quarter, washing the lifeboat out of its cradle, filling the main deck and tearing away the main hatch house which then went by the lee rail. 11 a.m. wind rounding to north moderated somewhat, all plain sail was set. Latitude at noon shows days run of 318 miles with speeds above 16 knots logged for 10 hours.*

Another entry in the Southern Ocean read, **SHIP'S LOG**: *4 p.m. a strong Northwest wind has increased to become a severe gale. All hands called to reduce sail. Topgallants, upper top-sails, courses and outer jibs taken in. 9 p.m. wind veered to west southwest and heavy seas boarding over quarters. The reefed fore coarse and upper fore topsail were reset, and the ship steered off the wind. At 10 p.m. a cresting sea pooped us, taking one of two helmsmen from the wheel and over the house to the main deck. Fortunately, he was little hurt.*

With her decks submerged, this ship raced on for days on end, logging 13 to 16 knots with her 450 miserable passengers battened below. Such was a sea passage to Australia in 1872.

By the beginning of the twentieth century the windjammers began to be built as really big ships with bridges, walkways fore and aft to keep men above the wave swept decks.

French four mast barque off the Horn wearing low sails.

Their masts, sails and yards were made the same size so as to be interchangeable. Soon the material of choice was steel to make them lighter and stronger. The crews of a windjammer consisted of about 14 to 20 men and in a moderate breeze, she sailed at speeds anywhere between 14 to 21 knots. The windjammers were used for those bulk commodities for whose transport, time was not a priority. Even in the great depression, some profit could be made in the grain trade, from Australia to Europe. The cargo rates varied from perhaps four dollars to eight dollars per ton and yes---that was difficult if you caught the low end of the rate. Still, the owners of one company purchased a ship for ten thousand dollars in 1931, and then loaded over 5,000 tons or 60,000 sacks of grain, for a gross income of $40,000. "The ship paid for herself and all her expenses for the year from the income of a single voyage even though half of it was in ballast. A general average for a ship's profit was $5,000 each a year after all expenses.

Archibald Russell casting off her tow, clearing for sea.

A crew had to love their work, for even a captain earned only a hundred a month in the 1930s while Ford Motors paid 25 per week on the assemble line.

Preussen was the largest sailing ship ever constructed. She was launched in 1902; the German five mast ship was huge, 485 ft length overall with a cargo capacity of 8,000 long tons and able to sail at over 20 knots. Big and tough the ship could not only sail in a hard gale, she could be maneuvered and tacked in one. Fast and able to weather any storm, *Preussen* earned the title of the *Queen of the Seas*. One record voyage was made, after another; and once she sailed from the Southern tip of England to Northern Chile in 57 days, faster than the steamers of her time.

Preussen

Sadly, on 5 November 1910, *Preussen* was rammed by the British cross-channel steamer *Brighten,* whose captain misjudged *Preussen's* 16-kt speed and attempted to cross her bow. In the collision, *Preussen* lost most of her forward rigging, including her bowsprit, and fore topgallant mast, which made steering the necessary course impossible. A November gale upset attempts to tow her into Dover and when her anchors carried away, *Preussen* went on the rocks at Crab Bay. *Preussen* was declared unsalvageable due to a broken back and her ribs protrude from the waters to this day.

Waiting to be scrapped.

The *Herzogin Cecile, Preussen* and the great sailors like her were the last of the mighty wind driven merchant ships. At the pinnacle of perfection, they disappeared from the earth in a decade. The very technology of steel construction and advanced methods of shipbuilding that made them possible, conspired with time, war and the worlds economics to make them extinct. So doomed, they passed from a substantial fleet to a few rusting museum ships. They now sail primarily as photographer's images or on the artists canvas. Never-the-less, they will always exist in the mind's eye for those with the love of beauty and a heart that can imagine.

Note: (A new *Herzogin Cecile* has been built, this time as a passenger cruise ship---if you'd like sailing experience aboard a square-rigger).

About the author

Roger Horton is a man who has worked at multiple professions but focused on four: merchant mariner, artist, author, and boat builder. He could add another thirty but they were merely skills practiced to pay bills. Asked what his true calling was, he'd tell you adventurer. Born in Toledo, Ohio in 1942, his family migrated south from the Great Lakes, settling in SE Florida in 1952. Florida had less than three million residents in the entire state back then. A kid's world was different in the 50's. Far less restrains were put on children; more responsibilities were given, and everybody worked. Like most boys then, he'd do whatever work was offered to earn a buck: mow lawns, wash windows, sell newspapers, even worked as a runner at Gulfstream racetrack. Slowly though, the jobs associated with boats, ships and the sea took hold. Marinas, boatyards, the seaport, had his interest. He'd skip school to sail on a breezy day; his parents were appalled. Roger was defiantly and without hope, a sea struck lad.

At 15, he went to sea (ran off actually). Initially sailed foreign flag, later with Windjammer Cruises, and then USMM. Enlisted in the Army in 1967, he served mainly in a marine transportation unit in Vietnam and was later an instructor at the US Army Marine Transportation School. On discharge from the Army in 1970 he returned to the USMM, sailing primarily as captain until retiring in 2005. During his career, he has sailed on about every type of vessel in the industry and a lot that were antiques, left behind by history. For fifty years he explored boatbuilding and studied the uses of watercraft on every continent.

Captain Horton also pursued parallel careers as novelist, marine artist and illustrator. His work can be found in numerous collections, and some work has been and is displayed in museums. Combined with his work as a marine artist, was research into the use of boats, and ships, and how they evolved throughout history. Presently, his time is divided between art, writing, and boating. For him boating includes, designing, building, modifying and of course sailing small yachts. Every year he plans on setting out on a few new adventures. Once this book is finished, he'll be off to sea again---where? Which way will the wind blow?

This book is dedicated to my children and now grandson who followed me to ships, boats and the sea as I followed my ancestors.

Preparing to take a Pilot aboard in lumpy weather.

Many a bold Pilot has lost his life while attempting to board a ship or while assisting a ship, crew and its passengers in emergencies. On the coast, inlets and rivers, the Pilot is the Captain's greatest asset.